Friederike Bathe

**The Role of Neck and Hinge for the Mechanism of Conventional Kinesins**

Friederike Bathe

# The Role of Neck and Hinge for the Mechanism of Conventional Kinesins

## A Functional Analysis of Fast Fungal Kinesin Motors

Südwestdeutscher Verlag für Hochschulschriften

**Impressum/Imprint (nur für Deutschland/ only for Germany)**
Bibliografische Information der Deutschen Nationalbibliothek: Die Deutsche Nationalbibliothek verzeichnet diese Publikation in der Deutschen Nationalbibliografie; detaillierte bibliografische Daten sind im Internet über http://dnb.d-nb.de abrufbar.

Alle in diesem Buch genannten Marken und Produktnamen unterliegen warenzeichen-, marken- oder patentrechtlichem Schutz bzw. sind Warenzeichen oder eingetragene Warenzeichen der jeweiligen Inhaber. Die Wiedergabe von Marken, Produktnamen, Gebrauchsnamen, Handelsnamen, Warenbezeichnungen u.s.w. in diesem Werk berechtigt auch ohne besondere Kennzeichnung nicht zu der Annahme, dass solche Namen im Sinne der Warenzeichen- und Markenschutzgesetzgebung als frei zu betrachten wären und daher von jedermann benutzt werden dürften.

Verlag: Südwestdeutscher Verlag für Hochschulschriften Aktiengesellschaft & Co. KG
Dudweiler Landstr. 99, 66123 Saarbrücken, Deutschland
Telefon +49 681 37 20 271-1, Telefax +49 681 37 20 271-0
Email: info@svh-verlag.de
Zugl.: Witten, Universität Witten/Herdecke, Diss., 2005

Herstellung in Deutschland:
Schaltungsdienst Lange o.H.G., Berlin
Books on Demand GmbH, Norderstedt
Reha GmbH, Saarbrücken
Amazon Distribution GmbH, Leipzig
**ISBN: 978-3-8381-1686-0**

**Imprint (only for USA, GB)**
Bibliographic information published by the Deutsche Nationalbibliothek: The Deutsche Nationalbibliothek lists this publication in the Deutsche Nationalbibliografie; detailed bibliographic data are available in the Internet at http://dnb.d-nb.de.

Any brand names and product names mentioned in this book are subject to trademark, brand or patent protection and are trademarks or registered trademarks of their respective holders. The use of brand names, product names, common names, trade names, product descriptions etc. even without a particular marking in this works is in no way to be construed to mean that such names may be regarded as unrestricted in respect of trademark and brand protection legislation and could thus be used by anyone.

Publisher: Südwestdeutscher Verlag für Hochschulschriften Aktiengesellschaft & Co. KG
Dudweiler Landstr. 99, 66123 Saarbrücken, Germany
Phone +49 681 37 20 271-1, Fax +49 681 37 20 271-0
Email: info@svh-verlag.de

Printed in the U.S.A.
Printed in the U.K. by (see last page)
**ISBN: 978-3-8381-1686-0**

Copyright © 2010 by the author and Südwestdeutscher Verlag für Hochschulschriften Aktiengesellschaft & Co. KG and licensors
All rights reserved. Saarbrücken 2010

# Contents

Abbreviations .................................................................................................................. V

1     **Introduction** .......................................................................................................... 1
    1.1     Molecular machines ....................................................................................... 1
        1.1.1     Cytoskeletal motor proteins ................................................................. 1
            1.1.1.1     Similarities to G-proteins ............................................................. 2
        1.1.2     The kinesin superfamily ...................................................................... 3
        1.1.3     Kinesin motors in filamentous fungi .................................................... 4
    1.2     Kinesin-1 from *N. crassa* ............................................................................. 5
        1.2.1     Functional anatomy .............................................................................. 5
            1.2.1.1     The minimal motor: head and neck-linker ................................... 6
            1.2.1.2     Optimizing motor function: the neck and hinge domain .............. 7
            1.2.1.3     Regulation and cargo binding: the stalk and tail domain ............. 8
        1.2.2     Motility model ..................................................................................... 9
            1.2.2.1     Hand-over-hand versus inchworm ............................................. 10
            1.2.2.2     Directionality ............................................................................. 10
        1.2.3     Characteristic features of NcKin ....................................................... 11
    1.3     Goal of the presented work ......................................................................... 11

2     **Materials and Methods** ..................................................................................... 13
    2.1     Materials ...................................................................................................... 13
        2.1.1     Reagents and other materials ............................................................. 13
        2.1.2     Vectors ............................................................................................... 13
        2.1.3     Bacterial strains ................................................................................. 13
        2.1.4     Media and cultivation of *E.coli* ......................................................... 14
    2.2     Molecular Biology Methods ....................................................................... 14
        2.2.1     Agarose gel electrophoresis ............................................................... 14
        2.2.2     DNA extraction from agarose gels ..................................................... 14
        2.2.3     Determination of DNA concentration ................................................ 15
        2.2.4     Preparation of plasmid DNA ............................................................. 15
        2.2.5     DNA cleavage with restriction endonucleases .................................. 15
        2.2.6     Ligation of DNA fragments into plasmid vectors ............................. 15
        2.2.7     Preparation and transformation of competent cells ........................... 15
            2.2.7.1     Preparation of electrocompetent cells ........................................ 15
            2.2.7.2     Electroporation ........................................................................... 16
            2.2.7.3     Preparation of SEM competent cells .......................................... 16
            2.2.7.4     Heat shock transformation ......................................................... 16

| | | |
|---|---|---|
| 2.2.7.5 | Analysis of transformed clones in *E. coli* | 16 |
| 2.2.8 | Polymerase chain reaction (PCR) | 17 |
| 2.2.9 | Point mutagenesis | 17 |
| 2.2.10 | Annealing of synthetic oligonucleotides | 18 |
| 2.2.11 | Oligonucleotides | 18 |
| 2.2.12 | Generation of constructs | 20 |
| 2.2.12.1 | Generation of Y362 point mutants | 20 |
| 2.2.12.2 | Generation of his-tagged, monomeric constructs | 21 |
| 2.2.12.3 | Generation of NcKin-HsKin chimeric constructs | 22 |
| 2.2.12.4 | Generation of pNcKin_stableNeck | 24 |
| 2.2.12.5 | Generation of chimeric constructs as 433-versions | 25 |
| 2.2.12.6 | Generation of P342C-constructs | 26 |
| 2.2.12.7 | Summary of all constructs used in this work | 26 |
| **2.3** | **Biochemical methods** | **29** |
| 2.3.1 | SDS-Polyacrylamide gel electrophoresis (SDS-PAGE) | 29 |
| 2.3.2 | Staining of SDS gels | 29 |
| 2.3.2.1 | Coomassie staining | 29 |
| 2.3.2.2 | Colloidal Coomassie staining | 30 |
| 2.3.3 | Expression of kinesin constructs | 30 |
| 2.3.4 | Protein purification | 30 |
| 2.3.4.1 | Chromatographic purification of bacterially expressed NcKin constructs | 30 |
| 2.3.4.2 | Chromatographic purification of bacterially expressed, his-tagged HsKin constructs | 31 |
| 2.3.4.3 | Affinity purification of kinesin constructs | 32 |
| 2.3.4.4 | Purification of pig brain tubulin | 33 |
| 2.3.5 | Determination of protein concentration | 34 |
| 2.3.6 | Determination of the oligomerization state | 34 |
| 2.3.6.1 | Sucrose density centrifugation | 34 |
| 2.3.6.2 | Gel filtration | 35 |
| 2.3.7 | Polymerization of microtubules | 35 |
| 2.3.8 | Determination of microtubule concentration | 36 |
| 2.3.9 | Fluorescent labelling of tubulin | 36 |
| 2.3.9.1 | Estimation of yield and labelling stoichiometry | 37 |
| 2.3.10 | Biotinylation of cys-tagged NcKin constructs | 38 |
| 2.3.11 | Multiple motor gliding assay | 38 |
| 2.3.11.1 | Video enhanced light microscopy | 38 |
| 2.3.11.2 | Gliding assay with biotin-labelled NcKin constructs | 39 |
| 2.3.11.3 | Gliding assay with hTail-constructs | 39 |
| 2.3.12 | Single molecule bead assay | 40 |
| 2.3.13 | ATPase measurements | 41 |
| 2.3.13.1 | Basal ATPase measurements | 41 |
| 2.3.13.2 | Microtubule-stimulated ATPase assay | 41 |

|     | 2.3.13.3 | Calculations for the coupled ATPase assay | 43 |
|     | 2.3.14 | Measurements under reducing and oxidizing conditions | 44 |
|     | 2.3.15 | MantADP release measurements | 44 |
|     | 2.3.15.1 | Generation of kinesin*mantADP complexes | 44 |
|     | 2.3.15.2 | Concentration of mantADP complexes | 45 |
|     | 2.3.15.3 | Stoichiometry of mantADP release | 45 |
|     | 2.3.15.4 | Pre steady-state kinetics of the mantADP release | 45 |

# 3 Results ........................................................................................................... 47

## 3.1 Sequence comparison of the neck and hinge domains of animal and fungal kinesins ........................................................................................................... 47

## 3.2 Characterization of point mutations in the neck domain ............................ 48
### 3.2.1 Design of the point mutants ........................................................................ 48
### 3.2.2 Isolation and biotinylation of NcKin constructs .......................................... 49
### 3.2.3 Motility ......................................................................................................... 50
### 3.2.4 ATPase measurements ................................................................................. 51
#### 3.2.4.1 Steady-state ATPase activity ...................................................................... 51
#### 3.2.4.2 Basal ATPase activity ................................................................................. 52
### 3.2.5 Oligomerization state ................................................................................... 53
### 3.2.6 MantADP release of NcKin-Y362K ............................................................ 56
#### 3.2.6.1 Stoichiometry of the microtubule-activated mantADP release ................... 56
#### 3.2.6.2 Pre steady-state kinetics of mantADP release ............................................ 57

## 3.3 Characterization of NcKin-HsKin chimeric mutants ................................... 59
### 3.3.1 Design of constructs .................................................................................... 59
### 3.3.2 Temperature dependence of NcKin and HsKin .......................................... 61
### 3.3.3 Gliding velocity and ATP turnover of chimeric constructs ........................ 63
#### 3.3.3.1 Motility ........................................................................................................ 63
#### 3.3.3.2 Microtubule-stimulated ATPase activity .................................................... 63
### 3.3.4 Oligomerization state ................................................................................... 65
### 3.3.5 Crosslinking studies on hinge chimeras ...................................................... 65
#### 3.3.5.1 Formation of crosslinks ............................................................................... 66
#### 3.3.5.2 Motility of crosslinked constructs ............................................................... 67
### 3.3.6 Characterization of a mutant NcKin with an artificial neck domain ........... 70
#### 3.3.6.1 Design of NcKin_stableNeck ...................................................................... 70
#### 3.3.6.2 Biochemical characterization of NcKin_stableNeck ................................... 71

## 3.4 Single molecule studies .................................................................................... 72
### 3.4.1 Affinity purification of motor constructs ..................................................... 72
### 3.4.2 Optical laser trap .......................................................................................... 73
#### 3.4.2.1 Principle ....................................................................................................... 73
#### 3.4.2.2 Experimental setup ...................................................................................... 74

      3.4.2.3    Sample preparation .................................................................................................. 76

   3.4.3    Observation of single NcKin molecules in the optical trap ............................................. 76

      3.4.3.1    Stall forces of single NcKin molecules ................................................................. 77

      3.4.3.2    Run length of single NcKin molecules under different external loads ................. 79

      3.4.3.3    Velocity of single kinesin molecules in the optical trap ....................................... 81

# 4    Discussion ........................................................................................................... 83

   4.1    Complexity of neck dimerization and implications for the regulation of fungal kinesins 83

   4.1.1    Functional importance of the conserved Tyr 362 ............................................................ 83

      4.1.1.1    Role of Tyr 362 for neck dimerization ................................................................. 84

      4.1.1.2    Regulatory function of Tyr 362 ........................................................................... 84

   4.1.2    Importance of the hinge region for the structural state of the NcKin neck domain ....... 87

      4.1.2.1    Role of the conserved Trp 384 for NcKin neck dimerization ............................... 87

      4.1.2.2    Evidence for intermediate structural states of the fungal neck domain ................. 87

      4.1.2.3    Integrity of the neck coiled-coil is required for NcKin motility ........................... 88

      4.1.2.4    Stabilization of the neck coiled-coil in NcKin ..................................................... 89

   4.1.3    Possible role of neck/hinge dynamics for the regulation of fungal kinesins ................... 91

   4.2    Role of the specific neck domain for NcKin motor mechanics ............................... 94

   4.2.1    The motor core determines fast ATPase and gliding activity in fungal kinesins ........... 94

      4.2.1.1    Temperature dependence of NcKin and HsKin ................................................... 94

      4.2.1.2    The specific fungal neck domain can be replaced by other kinesin sequences. .... 95

   4.2.2    Mechanic properties of an artifical coiled-coil in the fungal neck domain ................... 95

      4.2.2.1    Rationale for the design of the NcKin_stableNeck mutant .................................. 95

      4.2.2.2    The artificial neck satisfies basic requirements for NcKin motility ..................... 96

      4.2.2.3    Neck unwinding is not essential for NcKin stepping ........................................... 97

      4.2.2.4    Single molecule behaviour of NcKin wild-type .................................................. 98

      4.2.2.5    Single molecule behaviour of NcKin_stableNeck ............................................... 98

      4.2.2.6    Fine-tuning of NcKin motility by the neck domain ........................................... 100

   4.3    Conclusions and future prospects ........................................................................ 102

# 5    Summary ........................................................................................................... 104

# 6    References ......................................................................................................... 106

Acknowledgements ...................................................................................................... 116

# Abbreviations

| | |
|---|---|
| A | Arrhenius constant / Amplitude |
| ACES | N-[2-acetamido]-2-aminoethanesulfonic acid |
| ADP | adenosine-5'-diphosphate |
| AMP-PNP | adenosine-5'-[β,γ-imido]-triphosphate |
| AP100 | "Arbeitspuffer" with 100 mM PIPES |
| ATP | adenosine-5'-triphosphate |
| B | background activity |
| Bio | biotin |
| bp | base pairs |
| BRB80 | Brinkmann reconstitution buffer |
| BSA | bovine serum albumin |
| CD | circulardichroism |
| D | daltons |
| DNA | desoxyribonucleic acid |
| dNTP | desoxyribonucleotide triphosphate |
| DMSO | dimethylsulfoxide |
| DTNB | 5,5'-dithiobis(2-nitrobenzoic acid) |
| DTT | dithiothreitol |
| $\Delta E_A$ | Arrhenius activation energy |
| $E_x$ | extinction at wavelength x [nm] |
| ε | extinction coefficient |
| EDTA | ethylene-diamine-tetraacetic acid |
| EGTA | ethyleneglycol-bis-(2-aminoethylether)-N,N'-tetraacetic acid |
| EPR | electron paramagnetic resonance |
| FPLC | fast performance liquid chromatography |
| FRET | fluorescence resonance energy transfer |
| g | gravity |
| GFP | green fluorescence protein |
| GTP | guanosin-5'-triphosphate |
| $H_2O$ | distilled water |
| hktail / hTail | amino acids 342-546 of the tail domain of HsKin |
| HsKin | conventional kinesin of *Homo sapiens* |

| | |
|---|---|
| IPTG | isopropyl-β-thiogalactopyranoside |
| $k_0$ | basal ATPase rate |
| $k_{cat}$ | ATP turnover per second and head |
| $k_{max}$ | maximal rate constant |
| $k_{obs}$ | observed rate constant |
| $K_{D,MT}$ | dissociation constant for productive microtubule binding |
| $K_{0.5,MT}$ | half maximal activation constant |
| LDH | lactate dehydrogenase |
| M | molarity [mol/l] / seize marker |
| mantADP | 2'-(3')-0-[N-methylanthraniloyl]adenosine-5'-diphosphate |
| mantATP | 2'-(3')-0-[N-methylanthraniloyl]adenosine-5'-triphosphate |
| $M_r$ | molecular weight |
| mRNA | messenger ribonucleic acid |
| MT | microtubules |
| NADH | nicotine adenine dinucleotide |
| NcKin | conventional kinesin of *Neurospora crassa* |
| $OD_x$ | optical density at wavelength x [nm] |
| PAA | polyacrylamide |
| PAGE | polyacrylamide gel electrophoresis |
| PCR | polymerase chain reaction |
| PEP | phosphoenolpyruvate |
| pH | negative decadic logarithm of proton concentration |
| $P_i$ | phosphate |
| PI | protease inhibitor |
| PIPES | piperazine-N,N'-bis-[2-ethanesulfonic acid] |
| PK | pyruvate kinase |
| R | gas constant |
| rpm | rotations per minute |
| $r_{stokes}$ | Stokes-radius |
| $s_{w,20}$ | Svedberg constant |
| S | supernatant |
| F | "flow through" |
| SDS | sodium dodecyl sulfate |
| TIRF | total internal reflection fluorimetry |
| Tris | tris-hydroxymethyl-ammoniumethane |

| | |
|---|---|
| U | Units (enzyme activity) |
| $V_{max}$ | maximum rate of extinction decay |
| v | reaction rate |
| v/v | volume per volume |
| wt | wild-type |

Unless otherwise stated, SI-units, derived units and the decimal multiple of SI-units were used.

# 1 Introduction

## 1.1 Molecular machines

Molecular machines carry out a large variety of vital cellular processes. Some of these molecular machines, called motors, are specialized enzymes that generate mechanical work from another form of energy [Woehlke et al., 2000]. Basically these molecular machines are built up by one or more protein subunits but sometimes also contain nucleic acid components. Their specific cellular tasks range from DNA replication and transcription as carried out by DNA/RNA polymerases and helicases, RNA-translation by the ribosome and ATP generation by the $F_1/F_0$ ATP-synthase to cell movement by the bacterial flagellar motor, translocation by the nuclear pore complex and transport of cellular components by cytoskeletal motor proteins. The design principles differ greatly between different types of molecular machines that certainly did not evolve from a common origin. However, there are certain functional similarities that are shared by some motors that otherwise work in a fundamentally different way [Woehlke et al., 2000].
In the so-called rotary motors, a torque is generated by ion flux across the membrane that is coupled via a stator element to the rotary unit. The $F_1/F_0$ ATP-synthase or the much larger bacterial flagellar motor belong to this group of molecular motors. Other molecular machines have in common that they use some kind of molecular track, for example nucleic acids for DNA/RNA polymerases or actin and microtubule filaments for cytoskeletal motor proteins. Some of these machines are able to move for long distances along this track without detaching, a behaviour referred to as processivity [Schliwa et al., 2003].

### 1.1.1 Cytoskeletal motor proteins

The transport of many different cellular components to specific destinations within the cytoplasm and between membrane-bounded compartments is of fundamental importance for eukaryotic cells. The most widely used mechanism for intracellular transport involves a special class of molecular machines, the cytoskeletal motor proteins, often generally named molecular motors. They are designed to carry cargo directionally along a cytoskeletal track, some of them for long distances as single molecules. Three classes of cytoskeletal motors have evolved: the myosins, which use actin filaments as molecular tracks, and two classes of microtubule-dependent motors, the kinesins and the dyneins.

Cytoskeletal motor proteins undergo energy-dependent conformational changes that result in a directed movement along their filamentous tracks. They all possess a catalytic domain, also called motor domain or head, where ATP hydrolysis and binding of the filament take place. The size of the motor domain is very different in each class of molecular motor proteins. Kinesins have the smallest with about 350 amino acids, myosins possess an intermediate sized motor domain with about 800 amino acids and dyneins have the largest with over 4000 residues [Woehlke et al., 2000].

### 1.1.1.1 Similarities to G-proteins

The crystal structures of the catalytic domains of several myosins and kinesins are known and surprisingly revealed striking similarities although they virtually have no amino acid sequence identity and fundamentally different enzymatic properties [Kull et al., 1996]. To change between ATP and ADP-bound states, motor proteins must "sense" the presence or absence of the $\gamma$-phosphate group. In kinesins and myosins this is accomplished by conserved motifs, the so-called switch I and II loops that are almost identical in both motors and represent the $\gamma$-phosphate-sensor group [Kull et al., 1996; Woehlke, 2001]. Small conformational changes of the $\gamma$-phosphate-sensor driven by the cleavage of the ATP's phosphodiester bond are then transmitted to distant regions of the protein via the highly conserved switch II helix, linking the nucleotide state of the catalytic centre to the amplifying mechanical elements and the polymer binding site [Vale et al., 2000; Woehlke, 2001].

The nucleotide sensing motifs in the motor cores of kinesins and myosins are named in analogy to the catalytic centre in G-proteins. These proteins act as molecular switches that cycle between an activated GTP-bound state and an inactivated GDP-bound state, thereby controlling many cellular processes [Vetter et al., 2001]. The switch I and switch II regions were first identified within G-proteins, indicating that these motifs and the underlying molecular mechanism of adenine and guanine nucleoside-triphosphatases is ancient. Thus, G-proteins and the molecular motor families of kinesins and myosins most likely have a common ancestor protein [Kull et al., 1998; Vale et al., 2000; Vetter et al., 2001].

The dynein family of cytoskeletal motor proteins is fundamentally distinct from kinesins and myosins. The dynein motor head is a heptameric, ring-like structure containing a series of AAA domains (*A*TPases *a*ssociated with cellular *a*ctivities). ATPases of the AAA protein family are not related to G-proteins and do not contain the switch regions, Thus, the dynein motor class displays a completely different underlying mechanism than kinesins and myosins [Harrison et al., 2000].

## 1.1.2 The kinesin superfamily

In highly polarized structures like neuronal or epithelia cells, the long-range transport of organelles and vesicles is of fundamental importance. Thus, the first kinesin motor protein was not accidentally identified in the giant squid axon as motile force that underlies the movement of particles along microtubules [Brady, 1985; Vale *et al.*, 1985]. Since that time, a systematic molecular biological search has identified a fast-growing-superfamily of kinesins and kinesin-related proteins with more than ten subfamilies [Hirokawa, 1998; Schliwa, 2003; Schoch *et al.*, 2003; Dagenbach *et al.*, 2004], exhibiting a great variety of cellular functions – ranging from the transport of organelles, vesicles and mRNA throughout the cytoplasm to spindle assembly and integrity, chromosome motility, microtubule dynamics and trafficking of signalling modules (reviewed in [Hirokawa, 1998; Sharp *et al.*, 2000; Goldstein, 2001; Schnapp, 2003]). This functional versatility is basically located in the non-motor regions of the protein and is reflected by the large number of kinesins identified in the genomes of different organisms: 6 in *S. cerevisiae*, 24 in *Drosophila*, 33 in *C. elegans*, 45 in humans and 72 in *Arabidopsis* [Endow, 2003; Schoch *et al.*, 2003].

The classification of kinesin proteins is based on sequence similarities in the motor domain, containing highly conserved consensus motifs for ATP-binding and –hydrolysis and microtubule-binding. Very recently, a new nomenclature for all kinesin families has been proposed to clear confusions caused by many different nomenclatures based on diverse criteria and to facilitate communication among researchers [Lawrence, 2004].

The founding member of the kinesin superfamily is called kinesin-1 (conventional kinesin). Like most of the kinesins it possesses a N-terminally located motor domain and moves towards the plus end of the microtubule. Conventional kinesins are ubiquitously expressed in many tissues. They are primarily responsible for the plus end-directed transport of membranous organelles, like endosomes, lysosomes and vesicles in the endoplasmatic reticulum (ER) and the Golgi complex [Hirokawa, 1998]. Recently it was discovered that kinesin-1 also plays a role in positioning of signalling complexes in the cell, as indicated by the interaction with scaffolding proteins, like the JIP proteins that bind several members of the JNK/MAP kinase signalling pathway [Schnapp, 2003].

The monomeric members of the kinesin-3 family (previously referred to as Unc104/KIF1A family) are primarily responsible for the fast anterograde transport of presynaptic vesicles in neurons and the mitochondria transport [Hirokawa, 1998]. Other families like the kinesin-5 (former BimC-family) or the kinesin-4 play important roles in the assembly and maintaining of the mitotic spindle and the chromosome segregation [Sharp *et al.*, 2000]. The kinesin-2 family (heterotrimeric kinesins) consist of two different kinesin motor subunits that form a heterodimer and associate with a third, non-motor polypeptide. Their functions include ciliary and flagellar construction, for example the opsin transport in the photoreceptor-connecting cilium [Goldstein, 2001], as well as axonal transport in neuronal cells [Hirokawa, 1998].

Members of the kinesin-14 family possess a C-terminal motor domain and, in contrast to families with N-terminally located motor cores, they are minus-end directed motor proteins. They contain ATP-independent MT binding sites outside of the motor domain and therefore possess MT crosslinking activity. Many members of this family are involved in the spindle pole organization and act antagonistically to the plus-end directed kinesin-5 during pole segregation [Sharp *et al.*, 2000]. They also play a role in the retrograde transport and positioning of the Golgi-apparatus [Hirokawa, 1998; Ovechkina *et al.*, 2003].

Members of the kinesin-13 family possess an internally located motor domain. They differ greatly from all other kinesin families since they do not generate movement, but act as ATP-dependent MT depolymerases. They are primarily targeted to the MT ends where they induce or stabilize the curved GDP-state of the tubulin dimer at the MT end, leading to destabilization of the protofilament [Desai *et al.*, 1999; Hunter *et al.*, 2003]. Kinesin-13 members are located at kinetochores, centrosomes and also in the cytoplasm and nucleoplasm. By regulating MT dynamics they play a role in chromosome segregation, spindle elongation and are also thought to be involved in neurite extension [Ovechkina *et al.*, 2003].

### 1.1.3 Kinesin motors in filamentous fungi

Filamentous fungi grow in a highly polarized fashion to elongated hyphae, whereas the assembly of new cell wall material primarily occurs at the tip of the hyphae. This requires a distinct long-range transport within these organisms, which makes them an attractive model system to study intracellular transport in analogy to the axons in higher eukaryotes. Moreover, the growing extent of genome sequences and the ease of gene manipulation facilitate the identification of proteins involved in transport processes and their specific functions in fungi.

A complete inventory of kinesin motor proteins, based on consensus sequences within the motor domain, is available for several filamentous fungi, including *Neurospora crassa* [Schoch *et al.*, 2003]. Nine kinesin subfamilies have been found in filamentous fungi, with a total number of 10 kinesin proteins in *N. crassa*. They are involved in spindle formation and segregation, secretion and endocytosis, cell division, organelle and mRNA transport [Steinberg, 2000].

The first fungal conventional kinesin was isolated from *N. crassa* in 1995 [Steinberg *et al.*, 1995]. Since the kinesin null mutant is viable, in contrast to the situation in higher eukaryotes, a detailed cell biological characterization of this fungal motor was possible [Seiler *et al.*, 1997; Kirchner *et al.*, 1999a]. The deletion mutant showed severe defects in growth and cell morphogenesis, indicating an impaired transport of small, secretory vesicles carrying new cell wall material to the tip of the hyphae [Seiler *et al.*, 1997; Seiler *et al.*, 1999].

In the following years other fungal conventional kinesins have been found in *Syncephalastrum racemosum* [Grummt *et al.*, 1998a], *Ustilago maydis* [Lehmler *et al.*, 1997], *Nectria haematococca* [Wu *et al.*, 1998] and *Aspergillus nidulans* [Requena *et al.*, 2001]. All deletion mutants showed hyphal deformation and slower growth, but also additional phenotypes that vary in the different species. Vacuole formation and mitochondrial distribution were affected in kinesin null mutants of *N. haematococca* and *U. maydis* but not in *A. nidulans*, that instead showed defects in nuclear migration (for review: [Xiang *et al.*, 2004]. These results show that the same motor protein can be used for different cell biological functions in different organisms and that the participation of kinesin in cellular processes may be fine-tuned in a species-specific fashion.

## 1.2 Kinesin-1 from *N. crassa*

Sequence analysis of the kinesin-1 from *N. crassa*, NcKin, and other fungi compared to animal kinesin-1 showed striking similarities within the motor domain (55 % amino acid identity) but very little homology within the rest of the molecule. This clearly groups fungal kinesins to the kinesin-1 family, where they represent a distinct subgroup [Kirchner *et al.*, 1999b]. Accordingly, a detailed biochemical characterization of NcKin revealed strong similarities of the basic properties to animal conventional kinesins [Steinberg *et al.*, 1995; Steinberg *et al.*, 1996]. It has the same domain organization (see below), shows MT dependent ATPase activity and moves towards the plus-end of the microtubule.

### 1.2.1 Functional anatomy

On rotary shadow EM images kinesin-1 appeared as an elongated, rod-like structure, about 80 nm in length, with two globular heads (10 nm in diameter), followed by a stalk domain that is sometimes kinked in the middle, and a fan-shaped tail [Hirokawa *et al.*, 1989; Schliwa, 1989] (Fig. 1.1).

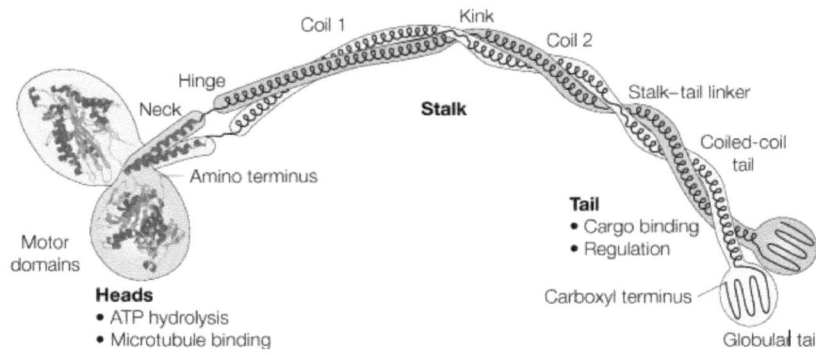

**Fig. 1.1: Domain structure of the conventional kinesin (kinesin-1) heavy chain dimer [from Woehlke & Schliwa, 2000].** The catalytic head (aa 1-332 in NcKin), the neck-linker (aa 333-341) and the coiled-coil neck domain (aa 342-374) show the crystal structure of the rat kinesin dimer [Kozielski et al., 1997]. The other domains are based on electron microscopic images [Hirokawa et al., 1989; Schliwa, 1989]. The hinge domain has an extended, flexible conformation [Seeberger et al., 2000]. The coiled-coil stalk region is divided into two subdomains by a flexible kink that allows the molecule to bend in the middle [de Cuevas et al., 1992]. The tail consists of a coiled-coil part that contains the cargo interaction site and a globular part, which is crucial for motor regulation [Kirchner et al., 1999a] [Seiler et al., 2000].

### 1.2.1.1 The minimal motor: head and neck-linker

The N-terminal motor domain or head comprises the first 332 amino acids in NcKin, with a similar size in animal kinesins. The crystal structure of the Nckin motor domain is known [Song et al., 2001] and reveales strong similarities with animal motor domain structures [Kull et al., 1996; Sack et al., 1997]. The central core contains an 8-stranded β-sheet and is flanked by 6 α-helices, 3 on each side [Song et al., 2001]. When free in solution, the kinesin motor domain is a very slow ATPase, but it is strongly activated (5500 x) upon binding to the microtubule [Kallipolitou et al., 2001]. Vice versa, the presence of ATP or ADP influences the binding affinity of the motor domain to the microtubule, having a high affinity in the nucleotide-free and ATP state and a weak affinity in the ADP state. For NcKin the affinity changes by a factor of 10 [Crevel et al., 1996; Crevel et al., 1999]. Thus, a communication pathway exists within the kinesin motor core that works in both directions between the nucleotide-binding pocket and the microtubule-binding region [Kull et al., 2002].

The conformational changes in the catalytic centre during ATP hydrolysis are propagated throughout the motor domain and amplified by further C-terminal regions of the motor to generate force and movement [Schliwa et al., 2003]. In kinesin-1, the neck-linker was shown to play an important role in co-operation with the motor core to produce motility. This short stretch of app. 10 amino acids (333 - 341 in Nckin) links the motor core to the start of the coiled-coil neck domain. The sequence of the neck-linker is strongly conserved in all conventional kinesins; however, there are two lysine residues in fungal kinesins that are not present in the animal kinesin sequence. In kinesin crystal structures, the neck-linker shows some conformational variability, being disordered in a human monomeric structure [Kull et al., 1996], but visible in rat kinesin monomeric and dimeric structures [Kozielski et al., 1997; Sack et al., 1997], another human kinesin structure [Sindelar et al., 2002] and in the NcKin structure, where it forms two short β-sheets that are separated by a kink. This transition from an unordered to an ordered state in different crystals structures suggests inherent mobility of the neck-linker, indicating a critical role for the motor mechanism. Confirming that view, replacement of the neck-linker with a random sequence results in a 200-250-fold decrease in gliding velocity [Case et al., 2000]. Moreover, a large conformational change of the neck-linker has been detected in human kinesin by cryo-EM and EPR techniques [Rice et al., 1999; Sindelar et al., 2002; Rice et al., 2003; Skiniotis et al., 2003]. In the nucleotide-free and ADP state, this domain is mobile but attaches to the motor core upon ATP binding, pointing towards the plus-end of the microtubule. Thus, the neck-linker is thought to serve as a mechanical amplifier that converts small conformational rearrangements within the motor core into motion [Vale et al., 2000; Schliwa et al., 2003].

**1.2.1.2 Optimizing motor function: the neck and hinge domain**

Although the motor core and the neck-linker represent the minimal motility-generating unit of kinesin-1, the subsequent neck and hinge domains are also required to create a fast and processive motor molecule [Woehlke et al., 2000].

The neck domain is the first part of the stalk (amino acids 342–374 in NcKin) that is predicted to form a coiled-coil structure as judged by the heptad repeats in the sequence with mostly hydrophobic residues in $a$ and $d$ positions [Huang et al., 1994b; Correia et al., 1995; Kallipolitou et al., 2001]. Peptide studies with synthetic peptides comprising the human kinesin neck region revealed the α-helical coiled-coil conformation [Morii et al., 1997; Tripet et al., 1997; DeLuca et al., 2001], also confirmed by crystal structures of the rat kinesin that showed a α-helical conformation in the first part of the neck in the monomer structure and a coiled-coil conformation in the structure of the dimer [Kozielski et al., 1997; Sack et al., 1997]. Initially, it was suggested that the kinesin neck might act as a lever arm analogue to the myosin α-helical tail. However, in the crystal structure the neck helix runs tangential to the microtubule surface rather then being in a 90° angle as it would be expected for a lever arm. Thus, in kinesin motors a different mechanism for force transduction is very likely [Inoue *et al.*, 1997; Sack *et al.*, 1997; Mazumdar *et al.*, 1998].

Nonetheless, the neck domain plays an important function for kinesin motility. Truncated constructs lacking the neck domain fail to dimerize, display greatly reduced gliding velocity [Berliner *et al.*, 1995] and do not move processively along the microtubule [Vale *et al.*, 1996b; Young *et al.*, 1998]. The monomer-dimer transition is also reflected in ATPase kinetics as the dimeric constructs show a much higher chemical processivity than monomers, indicating the hydrolysis of many ATP per encounter with the microtubule [Hackney, 1995; Jiang *et al.*, 1997b].

The hinge domain (amino acids 375–433 in NcKin) joins the motility generating motor core, neck-linker and neck to the coiled-coil stalk and tail domains. This region shows great variability in length and sequence and is not predicted to form a coiled-coil [Grummt *et al.*, 1998b]. Within the fungi, the region is rich in proline and glycine residues, suggesting an extended structure with high flexibility. The torsional freedom established by the hinge region may be important to enable the motor protein to align to the microtubule quickly in the right orientation [Hunt *et al.*, 1993]. However, the exact role of the hinge region remains elusive so far.

### 1.2.1.3 Regulation and cargo binding: the stalk and tail domain

The stalk domain of conventional kinesins is divided into two coiled-coil regions that are separated by a flexible kink [de Cuevas *et al.*, 1992]. A second flexible region links the stalk to the tail-coiled-coil, followed by the globular tail domain [Kirchner *et al.*, 1999a]. Sequence conservation is low throughout the stalk and tail regions, except for the last 30 amino acids of the second coiled-coil that are strongly conserved in animal kinesins [Kirchner *et al.*, 1999b]. This region is thought to represent the binding site of the light chains [Diefenbach *et al.*, 1998] that are involved in regulation of the motor and also mediate interaction with the cargo in animal kinesins [Verhey *et al.*, 2001].

Fungal kinesins lack light chains, indicating a more simple mechanism for regulation and cargo interaction that is located within the kinesin dimer itself. A folded conformation of the kinesin molecule has been detected by sucrose gradient sedimentation for both animal and fungal kinesins [Hackney *et al.*, 1992; Stock *et al.*, 1999; Seiler *et al.*, 2000], and was interpreted as an inactive state. The self-inhibition of the kinesin motor is mediated by an interaction of the globular tail domain, including the conserved IAK-motif, with the motor core and is thought to be relieved upon binding of cargo to the tail domain [Coy *et al.*, 1999; Friedman *et al.*, 1999; Stock *et al.*, 1999; Seiler *et al.*, 2000]. The putative site of cargo binding is located within the tail coiled-coil [Kirchner et al., 1999a; Seiler et al., 2000] and is strongly conserved within all kinesins, suggesting the same basic mechanism of cargo association in animals and fungi [Kirchner et al., 1999b].

## 1.2.2 Motility model

One of the most impressive properties of conventional kinesins is their remarkable processivity, which allows them to make several hundred steps along the microtubule before falling off. This enables the motor protein to transport the cargo over long distances up to several micrometers as single molecules [Lakamper et al., 2003]. Processive movement requires dimerization of the motor molecule or, more precisely, the presence of two heads [Hancock et al., 1998; Young et al., 1998; Hancock et al., 1999]. In a widely accepted model kinesin motors move in a hand-over-hand fashion [Cross, 2004] with the detachment of one head being contingent on the attachment of the other [Schief et al., 2001; Sablin et al., 2004].

As a prerequisite for processive movement at least one of the heads has to be bound to the microtubule at any time in the cycle. This is achieved by two mechanisms: i) the affinity of the motor head to the microtubule is changed in a nucleotide-dependent manner [Crevel et al., 1996] and ii) the hydrolysis cycles of both heads are kept enzymatically "out of phase" [Hackney, 2002], also called "alternating side catalysis" (Fig. 1.2) [Hackney, 1994a; Ma et al., 1997a; Gilbert et al., 1998]. When free in solution, both heads of the dimer are in the ADP bound state (0). Upon binding to the microtubule, the first head (in red) releases its ADP, whereas binding and ADP release of the second head (in green) is prevented (1) [Hackney, 1994a; Ma et al., 1997a]. Binding of ATP to the attached head causes a conformational change that allows the second, tethered head to bind to the next binding site in the plus direction on the microtubule and also to release its ADP (2). This state, also called the bridged state, where both heads of the kinesin motor are attached to the microtubule, is crucial for keeping the motor molecule on track for several cycles. However, in the crystal structure of the rat kinesin dimer, the two heads are not capable to span the 8 nm distance between two adjacent tubulin binding sites [Kozielski et al., 1997]. Thus, major conformational changes have to occur to allow the bridged conformation. In a widely accepted view, the rearrangement of the flexible neck-linker domain, as discussed above, accommodates simultaneous binding of the heads [Romberg et al., 1998; Tomishige et al., 2000].

The intramolecular strain that is generated in the bridged state is thought to prevent ATP binding to the leading head before the trailing head has hydrolysed ATP and released the Pi [Rosenfeld et al., 2003]. In the ADP-state, the microtubule affinity is weakened, so the trailing head detaches from the track while the leading head holds on (3). At this point the two heads of the dimer have changed their roles and the cycle starts again.

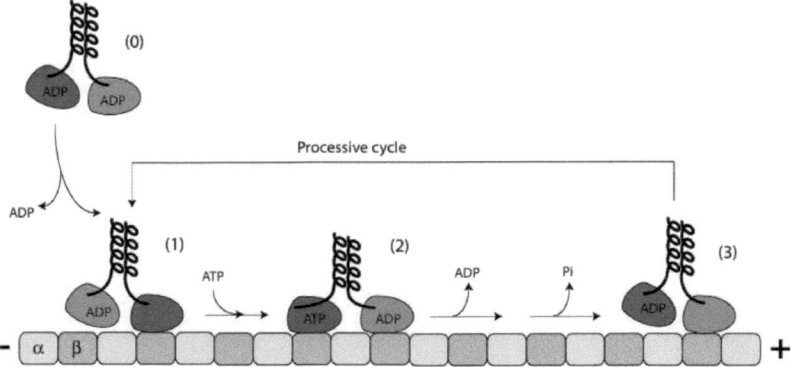

**Fig. 2: Processive movement of kinesin in the hand-over-hand mechanism.**
Processive movement requires two heads that are strictly coupled to undergo alternating site catalysis. Microtubule affinity is changed in a nucleotide-dependent fashion, so at least one of the heads is in a strong-binding state, containing either ATP or no nucleotide.
Kinesin constructs are depicted only up to the coiled-coil neck domain. Tubulin α subunits are highlighted in yellow, β subunits in orange. The catalytic head domains are coloured differently to illustrate their alternating relative positions. For explanations see text.

### 1.2.2.1 Hand-over-hand versus inchworm

Models other than the hand-over-hand mechanism are discussed for kinesin motility. The "inchworm" model for processive movement proposes that the two heads of the kinesin dimer retain their non-equivalent positions to each other, with one head being always the leading head and one head always the trailing head [Hua et al., 2002]. However, measuring the stepwise movement of single motors, particularly the distribution of dwell times of consecutive steps [Asbury et al., 2003; Kaseda et al., 2003] and the exact position of one head during stepping [Yildiz et al., 2004] provided strong evidence for an asymmetric hand-over-hand mechanism.

### 1.2.2.2 Directionality

All minus-end directed kinesins studied so far possess a C-terminally located motor domain (kinesin-14 family, 1.1.2.2). Chimeric constructs, swapping the motor domains of forward and reverse motors, as well as mutational studies revealed the head-neck interaction to be the critical factor in determining directionality [Henningsen et al., 1997; Sablin et al., 1998; Endow et al.,

2000]. Apparently, the kinesin motor core possesses an intrinsic bias for plus-end directed motility that is rectified by the neck-linker and neck domain in conventional kinesins to generate plus end-movement [Case et al., 2000]. Minus-end directed motors like ncd have no neck-linker domain. Most likely, direct interactions of the conserved neck helix of ncd with the motor core revert the directional bias of the head domain and force the molecule to move forward.

### 1.2.3 Characteristic features of NcKin

Although NcKin shares many important features with animal kinesin-1, as stated above, it also exhibits several unique and functionally significant properties.

For example, NcKin was shown to be a homodimer, consisting of two identical heavy chains [Steinberg et al., 1996], whereas animal kinesin-1 is a heterotetramer with two heavy chains (KHC, 110-150 kD) and two light chains (60-80 kD) [Verhey et al., 1998]. Moreover, the head-tail inhibition is less pronounced and much more salt-sensitive in NcKin compared to animal kinesin-1 [Seiler et al., 2000; Kallipolitou et al., 2001]. These findings suggest different regulatory mechanisms for kinesin-1 in fungi and animals.

The most characteristic feature of fungal kinesins is their high *in vitro* microtubule gliding velocity: with 2.0-2.8 $\mu$m/s they move 4-5 fold faster than human conventional kinesins [Steinberg et al., 1995; Grummt et al., 1998a]. As in animal kinesin-1, motile and kinetic characteristics clearly depend on the oligomerization state, as shown with C-terminally truncated NcKin constructs [Kallipolitou et al., 2001]. However, NcKin exhibits an unusual dimerization behaviour. Whereas in animal kinesins the neck domain induces the formation of stable dimers, the fungal neck domain is not sufficient for dimerization and requires additional motifs from the hinge domain. The presence of an unfolded neck domain in these so-called "long monomeric" NcKin constructs leads to the inhibition of the ATPase activity, indicating a direct influence of the neck domain on the catalytic motor core. Consistent with their unique features, fungal neck domains possess a strictly conserved sequence pattern, clearly different from animal kinesin-1.

## 1.3 Goal of the presented work

As stated above, the tight connection between the two heads of a kinesin motor, provided by a stable neck coiled-coil, is essential for the communication between both heads and thus for processive movement. However, the function of the kinesin neck region is probably not simply confined to dimerization.

In some studies, transient melting of the neck coiled-coil during kinesin stepping is proposed to allow for spanning the 8 nm step size [Tripet *et al.*, 1997; Hoenger *et al.*, 1998; Hoenger *et al.*, 2000]. This was rendered unlikely by other studies that replaced the neck with a stable coiled-coil sequence [Romberg *et al.*, 1998] or that introduced a crosslink at the N-terminus of the neck [Tomishige *et al.*, 2000]. However, the insertion of an artificial neck sequence mildly but significantly reduces processivity [Romberg *et al.*, 1998]. Moreover, deletion or insertion of charged residues in the neck domain also severely changes processive behaviour of the motor [Thorn *et al.*, 2000], indicating an important role of the specific neck sequence for kinesin mechanochemistry, beyond dimerization.

The functional role of the hinge region is even more enigmatic. Deletion of the hinge in a fungal kinesin severely reduces the gliding velocity but accelerates ATP turnover, indicating an important function for the coupling of ATP hydrolysis and movement [Grummt *et al.*, 1998b]. However, sequence analysis reveals very poor conservation within the hinge region of all conventional kinesins. Furthermore, the hinge region is not predicted to form a coiled-coil and displays a high degree of flexibility [Seeberger *et al.*, 2000], as indicated by a high content of proline and glycine residues in the fungal sequence. However, the replacement with an artificial, proline-rich sequence could not rescue the phenotype of the deletion construct [Grummt *et al.*, 1998b], suggesting a distinct function of the wild-type sequences.

The goal of the present work was to characterize the functional role of the neck and hinge domains in kinesin-1 in more detail. The fungal kinesin-1 NcKin was used as a model system, since fungal kinesins are "natural mutants" that share their basic mechanism with animal conventional kinesins, but display some unique features, which may represent a fine-tuning of the kinesin motor to accomplish specific demands. To investigate their specific functions either single, conserved residues or the entire neck and/or hinge region of NcKin were replaced by human kinesin or artificial sequences. The resulting mutants were characterized using oligomerization studies, motility assays, steady-state and pre steady-state kinetics and single-molecule techniques.

# 2 Materials and Methods

## 2.1 Materials

### 2.1.1 Reagents and other materials

Unless otherwise stated chemicals were obtained from Biorad (München), Fluka (Buchs, Schweiz), Merck (Darmstadt), Roche-Diagnostics (Penzberg), Carl Roth (Karlsruhe), Serva (Heidelberg) and Sigma-Aldrich (Deisenhofen) and were of p. a. quality. Other consumables were mainly supplied from Greiner (Frickenhausen), Nunc (Wiesbaden), Qiagen (Hilden) and Sarstedt (Nümbrecht).

### 2.1.2 Vectors

For *Neurospora crassa* kinesin and chimeric constructs:
pNK433cys [Kallipolitou *et al.*, 2001]
pNk433_C307A_hkTail [Hahlen, 2004a]
based on pT7-7 (Tabor)

For human kinesin constructs:
pHK546 [Vale *et al.*, 1996b]
pHK546_$N_2K_2$ [Kallipolitou, 2002]
based on pET-17b (Novagen, Madison)

### 2.1.3 Bacterial strains

*Escherichia coli* strains DH5α [Sambrook *et al.* 1989] and XL1-Blue (Stratagene, Amsterdam) were used for cloning. *E. coli* strain BL21 CodonPlus (DE)-RIL [Studier *et al.* 1990] (Stratagene) was used for protein expression.

### 2.1.4 Media and cultivation of *E.coli*

*E. coli* cells were grown according to standard methods [Sambrook *et al*. 1989] on agar plates or shaking cultures (240 rpm, 37°C). For protein expression the temperature was reduced to 22°C.

Media:

| | |
|---|---|
| LB: | 1% tryptone, 0.5 % yeast extract, 0.5% NaCl |
| SOB: | 2% tryptone, 0.5% yeast extract, 10 mM NaCl, 2.55 mM KCl |
| Agar plates: | 1.5% agar in LB medium |
| TPM: | 2% bacto-tryptone, 1.5% yeast extract, 0.8% NaCl, 0.2% $Na_2HPO_4$, 0.1% $KH_2PO_4$, after autoclaving: + 0.2% glucose |
| Antibiotics: | after autoclaving: |
| | Ampicillin:  100 µg/ml |
| | Chloramphenicol:  25 µg/ml |

## 2.2 Molecular Biology Methods

### 2.2.1 Agarose gel electrophoresis

The separation of DNA fragments according to their size was performed using gels with 0.8% to 2% agarose in TAE buffer. For detection of DNA fragments 0.05 µg/ml ethidium bromide was added to the liquid agarose. Before loading samples were mixed with DNA loading dye. Gels were run with 75 V. Bands were detected under UV illumination and documented using the Eagle Eye II CCD camera system (Stratagene, Heidelberg).

| | |
|---|---|
| 50x TAE: | 2 M Tris-HCl, 0.57% acetic acid, 50 mM EDTA; pH 7.5 |
| 6x DNA loading dye: | 30% glycerol, 0.25% bromphenol blue, 0.25% xylene cyanol |

### 2.2.2 DNA extraction from agarose gels

DNA bands were excised with a scalpel, transferred to sterile Eppendorf vials and purified using the "Qiaquick gel extraction kit" (Qiagen, Hilden), following the instructions of the manufacturer. Finally, DNA was resolved in 30 µl $H_2O$.

## 2.2.3 Determination of DNA concentration

DNA concentration in solutions was determined photometrically at 260 nm ($E_{260}$), with 50 µg/ml of double-stranded DNA corresponding to $E_{260}$ of 1.0 [Sambrook et al. 1989].

## 2.2.4 Preparation of plasmid DNA

Plasmid DNA was prepared from overnight shaking cultures using the Qiagen-Plasmid-Kit (Qiagen, Hilden). For analytical scale preparations (3 ml) the manufacturer's manual for "mini-preps" excluding the Tip20-column was followed, for larger scale preparations (100-200 ml) the manual for "midi-preps".

## 2.2.5 DNA cleavage with restriction endonucleases

Restriction enzymes were obtained from Roche-Diagnostics (Penzberg) and New England Biolabs (Schwalbach). Restriction digests were performed using the buffer and temperature conditions recommended by the manufacturer. 1-5 units enzyme were used per µg DNA. The reaction volume was at least 15 µl, including 1 mg/ml casein. Samples were incubated for 1.5-3 h and analysed on agarose gels (2.2.1).

## 2.2.6 Ligation of DNA fragments into plasmid vectors

Vector and DNA fragments were cleaved (2.2.5), separated on agarose gels by electrophoresis (2.2.1), and extracted from agarose gels (2.2.2). DNA fragments were ligated with T4 DNA ligase (New England Biolabs, Frankfurt) in a volume of 15 µl at 16°C for at least 2 h using the buffer system supplied by the manufacturer. 50 ng of vector were incubated with no insert (religation control), equimolar amount and a three- to sixfold molar excess of insert. 5 µl of the sample was transformed into competent E. coli cells (2.2.7.2 and 2.2.7.4).

## 2.2.7 Preparation and transformation of competent cells

### 2.2.7.1 Preparation of electrocompetent cells

1 l LB medium was inoculated with 10 ml of an E. coli overnight culture and grown to an $OD_{600}$ of 0.6 at 37°C under vigorous shaking. All flasks and solutions subsequently used were sterilised and cooled to 4°C. Quality of the competent cells depended on consequent chilling. Cells were

harvested by centrifugation (GSA rotor: 4000 rpm, 15 min, 4°C) and resuspended in 1 l $H_2O$. After another centrifugation the cells were resuspended in 500 ml $H_2O$, pelleted again, washed with 20 ml of 10% glycerol and finally resuspended in 3 ml of 10% glycerol. After aliquotation the cells were frozen in liquid nitrogen and stored at -70°C.

#### 2.2.7.2 Electroporation

For transformation, electrocompetent cells were thawed on ice. 50 µl cells were mixed with 0.5 µl vector or 5 µl ligation reaction and placed in a pre-cooled, sterile electroporation cuvette (Eurogentec; distance between electrodes 2 mm). After a pulse (2.5 kV, 25 mF) the whole cell suspension was plated on LB agar plates with 100 µg/ml ampicillin (DH5α and XL1-Blue) or with 100 µg/ml ampicillin and 25 µg/ml chloramphenicol (BL21-RIL).

#### 2.2.7.3 Preparation of SEM competent cells

250 ml SOB medium was inoculated with 3 ml of an *E. coli* overnight culture and grown to an $OD_{600}$ of 0.6 at 37°C under vigorous shaking. After the culture was incubated on ice for 10 min the cells were harvested by centrifugation (GSA rotor: 2500 rpm, 10 min, 4°C) and resuspended in icecold TB solution. After another centrifugation the cells were resuspended in 20 ml TB and 7 % DMSO (v/v) was added to the cell suspension. The cells were frozen in small aliquots in liquid nitrogen and stored at –70 °C [Inoue *et al.*, 1990].

TB solution: 10 mM PIPES•KOH, 55 mM $MnCl_2$, 15 mM $CaCl_2$, 250 mM KCl, pH 6.7

#### 2.2.7.4 Heat shock transformation

For transformation, electrocompetent cells were thawed on ice. 50 µl cells were mixed with 0.5 µl vector or 5 µl ligation reaction and incubated on ice for 30 min. The mixture was placed in a 42°C waterbath for 30 s, then on ice for 2 min and was plated immediately on LB agar plates with 100 µg/ml ampicillin (DH5α and XL1-Blue) or with 100 µg/ml ampicillin and 25 µg/ml chloramphenicol (BL21-RIL).

#### 2.2.7.5 Analysis of transformed clones in *E. coli*

DNA of transformed bacteria was isolated (2.2.4), cleaved with appropriate restriction endonucleases (2.2.5) and analysed on agarose gels (2.2.1). Plasmids with restriction fragments of the expected size were sequenced by Medigenomix, Martinsried or Biolux, Stuttgart. Sequences were analysed using the GCG software (Wisconsin package, GCG Inc.).

## 2.2.8 Polymerase chain reaction (PCR)

Amplification of DNA fragments was carried out by the polymerase chain reaction.

The "Expand High Fidelity Polymerase Mix" (Roche, Penzberg) was used for standard PCR reactions. The reactions contained 2 mM $MgCl_2$, 2 ng/$\mu$l template (final concentration), 200 $\mu$M of each dNTP, 0.5 $\mu$M 5'- and 3'-primer and 0.5 units per 50 $\mu$l expand-polymerase in reaction buffer. Plasmid DNA was used as template. Number of cycles, temperature and duration of denaturation, annealing and elongation phases were chosen according to the supplier's instructions. The PCR product was isolated from nucleotides and enzyme by the "Qiaquick PCR Purification Kit" (Qiagen, Hilden).

The "QuikChange" (Stratagene Cloning Systems) PCR protocol was used to introduce point mutations (2.2.9). The reactions contained 2 mM $MgCl_2$, 0.1-1 ng/$\mu$l template (final concentration), 50 $\mu$M of each dNTP, 0.2 $\mu$M 5'- and 3'-primer and 2.5 units per 50 $\mu$l *Pwo*-polymerase (without exonuclease activity) in reaction buffer.

For the screening of large numbers of clones, analytical "colony PCR" reactions were performed. The reaction mix was similar to the standard PCR protocol described above, however no plasmid DNA but whole cells, freshly picked from a transformation plate, were used as template. The primer were chosen to anneal within the desired insert, so positive clones could be identified with the presence of PCR product. The cheaper *Pfu*-DNA-polymerase (provided by Ralph Gräf) was used instead of the "Expand Polymerase" for analytical PCR reactions. To promote cell lysis, the denaturation step in the first cycle was 10 min at 95 °C instead of 2 min for other protocols.

## 2.2.9 Point mutagenesis

Mutations that lay close to restriction sites of the vector were introduced by designing primers that contain an original restriction site of the vector and the intended mutation. The PCR product was purified, cleaved with the appropriate restriction enzymes and ligated into the vector.

If there were no suitable restriction sites near a desired mutation site, the "QuikChange" (Stratagene Cloning Systems) mutational protocol was used. Complementary primers containing the mutation and the *Pwo*-polymerase without 3' – 5' exonuclease activity were used (2.2.8). To remove the bacterial template DNA not containing the intended mutation, the PCR mixture was cleaved with *Dpn*I, which is specific for methylated DNA. The PCR product was then transformed in *E. coli* cells. The DNA was prepared, a minimal fragment containing the mutation was cut out by restriction endonucleases and re-ligated into the vector to remove undesired mutations introduced by the *Pwo*-polymerase.

All clones containing introduced mutations were verified by sequencing.

## 2.2.10 Annealing of synthetic oligonucleotides

For the annealing of complementary oligonucleotides, 10 $\mu$l of the + and the – oligo with a concentration of 10 pmol/$\mu$l were mixed and supplemented with 1 x PCR-buffer ("Expand High Fidelity Polymerase buffer", Roche, Penzberg). The mixture was heated for 5 min at 95°C and cooled down slowly to room temperature (1-2 h) to promote specific annealing.

If the oligonucleotides were not completely overlapping, the single-stranded parts were filled up by adding 0.5 $\mu$l Expand-polymerase and 2 $\mu$l dNTP's (0.2 mM each) and incubated at 72°C for 5 min.

## 2.2.11 Oligonucleotides

Oligonucleotides were purchased from ThermoHybaid (Ulm). The sequences of the following oligonucleotides are given from 5' to 3':

**PCR-primer for the introduction of point mutations**
**Y362F+ :**
CGTCCTTCGAGAACTTTATCGTCAACCTGGCCA
**Y362F- :**
CCAGGTTGACGATAAAGTTCTCGAAGGACGCGT
Complementary primer for site-directed mutagenesis.

**PCR-primer for the generation of monomeric constructs**
**pT7.7+ :**
CTCACTATAGGGAGACCACAACG
Universal pT7.7 primer, annealing in the T7-promotor

**NK344Pst-**
AACTGCAGCTCGGCCGGGCTGAGTTCGGC
Generation of NcKin344; introduction of *Pst*I-site

**hk339-**

AACTGCAGCTGTTCAGCAGTTAACTCCACATTGAC

Generation of pHK339; introduction of a *Pst*I-site

**PCR-primer for the generation of chimeric constructs**

**HK560(Pst)- :**

AACTGCAGTTAAAAGCATTTACGATGCACAATAGACGG

Generation of pNK433_379_hktail, pNK433_384_hktail and pNK_340_hktail; introduction of a *Pst*I-site

**HK560(Kpn)+ :**

CGTAATGGGGAGACGGTACCTATTGATGAACAG

Generation of pNK433_379_hktail and pNK433_340_hktail; introduction of a *Kpn*I-site

**HK560(KpnIW384)+ :**

GGGGTACCCAAGGAGAAATGGTTTGACAAAGAGAAAGCCAACTTGG

Generation of pNK433_384_hktail; encoding NcKin 380-84; introduction of a *Kpn*I-site

**HK560(KpnIΔSacI)- :**

GGGGTACCGTCTCCCCATTACGCCATCTGTTGAGTTCATTTTCAAGC

Generation of pNK433_340-79_hktail; introduction of a *Kpn*I-site and removal of a *Sac*I-site (772 in pHK546)

**HK560(SacIΔPstI)+ :**

GCGGCGAGCTCACCGCAGAACAGTGG

Generation of pNK433_340-79_hktail; introduction of a *Sac*I-site and removal of a *Pst*I-site (856 in pHK546)

**Oligonucleotides for the introduction of tags**

**His8+ :**

GCATCACCATCACCATCACCATCACTA

**His8- :**

AGCTTAGTGATGGTGATGGTGATGGTGATGCTGCA

Cloning of his8-tag into pNK433cys; *Pst*I and *Hind*III compatible

**oCystagNgoMIV+ :**

CCGGCACTCCCAGCTTGCCCTTGGACCCGTCTATTGTGCATCGTAAATGCTTTTAACTGCA

**oCystagNgoMIV- :**

GTTAAAAGCATTTACGATGCACAATAGACGGGTCCAAGGGCAAGCTGGGAGTG

Cloning of cys-tag into pNK433_C307A_hktail, after removal of the hTail; *NgoM*IV and *Pst*I compatible

**Oligonucleotides for the generation of pNcKin_stableNeck**

**coiled oligo+ :**

ATCGAAGCTTTGAAAGCTGAAATCGAAGCTTTGAAAGCTGAAATTGAGGCGCTCAAGGCGGAG

**coiled oligo- :**

TTTCAAAGCTTCGATTTCAGCTTTCAAAGCTTCGATCTCCGCCTTGAGCGCCTCAAT

Cloning of the artificial EIEALKA (stableNeck) sequence

**stable coil+ :**

AACGCCGAGCTCAGCATCGAAGCTTTGAAAGCTGAA

**stable coil- :**

GGGGTACCGTCTCGCCACCTTTCAAAGCTTCGAT

PCR-primer for amplification of the stableNeck sequence; introduction of *Kpn*I and *Sac*I restriction sites

For oligonucleotides that were used for sequencing, see [Kallipolitou, 2002; Hahlen, 2004a].

### 2.2.12 Generation of constructs

#### 2.2.12.1 Generation of Y362 point mutants

All Y362 point mutants are in the background of the parent plasmid **pNK433cys**, cloned by Athina Kallipolitou (2001). This plasmid encodes a C-terminally truncated version of the NcKin protein, containing the first 433 amino acids up to the hinge domain (NcKin$_{433}$). C-terminal to the construct there is an appendix of 9 amino acids, called the "cys-tag", containing a reactive cysteine residue that can be easily labelled (PSIVHRKCF, [Funatsu *et al.*, 1997]).

The point mutation Y362K was originally introduced by Ulrike Majdic (1999) in the NcKin full length construct (pNK928-Y362K) and subcloned via PCR into the pNK433cys vector by Günther Woehlke (**pNK433cys-Y362K**). The point mutations **Y362C** and **F369C** were introduced into the plasmids pNK433-Y362C_light and pNK433-F359C_light by Manuela Ludwig. In the so-called "cys-light" plasmids all three native cysteines were exchanged into alanine or serine (C38S, C59S, C307A) [Hahlen, 2004a]. Since constructs containing the C38S/C59S double exchange could not be expressed in *E. coli*, the sequences for NcKin$_{433}$-F359C and NcKin$_{433}$-Y362C were subcloned into the pNK433cys background, using the restriction sites *Pvu*I (within the Ampicillin resistence gene of the vector) and *Bam*HI (within the NcKin head). The resulting fragments (2.1 kB) contained the mutated codons and were introduced into the pNK433cys vector that had been cut using the same enzymes. Since the "cys-light" plasmids contain no cys-tag and the tag of the pNK433cys vector was removed during the *Pvu*I/ *Bam*HI double digestion, the resulting pNK433-Y362C and –F359C plasmids does not contain a "cys-tag".

The **Y362F** point mutation was introduced into the pNK433cys vector via site-directed mutagenesis with the QuikChange protocol (Stratagene Cloning Systems) (2.2.9), using the oligonucleotides Y362F+ and Y362F- (2.2.10). A 780 bp fragment containing the Y362F mutation was back-cloned into the NK433cys vector via the restricition sites *Sac*II and *Pst*I.

All clones used for protein expressions were verified by sequencing.

### 2.2.12.2 Generation of his-tagged, monomeric constructs

To improve the purification procedure of truncated constructs, a "his8-tag" consisting of 1 glutamine residue and 8 histidine residues (-QH$_8$) was introduced C-terminal to the constructs.

In a first step, the his-tag was introduced into the pNK433cys vector (pNK433his). Two oligonucleotides that encode the his-tag and a stop-codon were constructed, being complementary to each other and *Pst*I-compatible at the 5'end and *Hind*III-compatible at the 3' end (referring to the sense-strand). The oligonucleotides were annealed, resulting in a double-stranded, artificial DNA fragment, and then ligated into the pNK433cys vector that had been cut with *Hind*III and *Pst*I. Since the stop-codon lies upstream the *Pst*I-site, referring to the reading direction, the his-tag is not part of the NcKin-gene in the pNK433his plasmid. This construct was used to subsequently generate pNK344his and pHK339his.

The plasmid **pNK344his** encoding a truncated, his-tagged NcKin construct, was cloned by PCR on the parent plasmid pNK433cys using the oligonucleotides NK344Pst- and pT7.7+ as primer. The NK344Pst oligonucleotide introduces a *Pst*I restriction site immediately behind the codon for aa 344. The amplified fragment was cleaved with *Pst*I and *Xba*I and the resulting fragment (1034 bp) was ligated into the pNK433his plasmid that has been cut using the same enzymes. The resulting plasmid pNK344his encodes for NcKin$_{344}$ including the C-terminal his-tag.

The plasmid **pHK339his** encoding a truncated human kinsin construct with a his-tag was also cloned via PCR using the oligonucleotides hk339- and pT7.7+ as primer. As PCR template the plasmid pHK546_N$_2$K$_2$, constructed by Athina Kallipolitou [Kallipolitou, 2002], was used. This plasmid is based on the parent construct pHK546 with an additional *Ngo*MIV-site between the hinge and the tail domain (nucleotide position 603) and a *Kpn*I-site between the neck and the hinge domain (nucleotide position 735). The hk339- oligonucleotide introduces a *Pst*I restriction site behind the codon for aa 339. The amplified fragment was again cleaved with *Pst*I and *Xba*I, the resulting fragment (1017 bp) was then ligated into the pNK433his plasmid that had been cut using the same enzymes. The resulting plasmid pHK339his encodes for HsKin$_{339}$ including the C-terminal his-tag.

All clones used for protein expressions were verified by sequencing.

### 2.2.12.3 Generation of NcKin-HsKin chimeric constructs

All chimeric constructs were cloned into the background of the **pNK433_C307A_hktail** parent plasmid constructed by Katrin Hahlen [Hahlen, 2004a]. This plasmid contains a changed codon (C307A) and the following restriction sites corresponding to the domain borders of the NcKin protein: a *Ngo*MIV-site at 425 (close to the C-terminal end of the hinge domain), an unique *Kpn*I-site (native *Kpn*I site deleted) at 378 (between the neck and the hinge domain) and a *Sac*I-site at 339 (between the neck and the neck-linker). Moreover, the hTail domain, containing NcKin to amino acid 436 followed by amino acids 432 to 546 in the HsKin human kinesin sequence, was cloned via *Pst*I and *Ngo*MIV into the pNK433_C307A plasmid, resulting in pNK433_C307A_hktail.

For generation of chimeric constructs, the desired HsKin sequences were amplified via PCR on the pHK546_N$_2$K$_2$ plasmid, using the appropriate primer that introduced the corresponding restriction sites, and cloned into the pNK433_C307A_hktail vector. As an example, cloning of **pNK433_379_hktail** (protein: NcKin_379$_{hTail}$), comprising the NcKin head, neck linker and neck sequences up to aa 379 (NcKin numbering) and the HsKin hinge and tail sequences from 375 to 546 (HsKin numbering) is shown in more detail in fig. 2.1.

**pNK433_379_hktail:** PCR on pHK546_N$_2$K$_2$ with HK560(Kpn)+ and HK560(Pst)-, introducing a *Pst*I-site at the C-terminus of the protein (nucleotide position 153 in the human kinesin sequence) and a *Kpn*I-site at the neck/hinge domain border. Cloning via *Pst*I and *Kpn*I into pNK433_C307A_hktail.

**pNK_384_hktail:** PCR on pHK546_ N$_2$K$_2$ with HK560(Pst)- and HK560(KpnIW384)+, the latter introducing a *Kpn*I-site and codons for NcKin residues 380-384. Cloning via *Pst*I and *Kpn*I into pNK433_C307A_hktail. The resulting construct contains the NcKin head, neck-linker and neck domain including the first 5 hinge residues up to W384. The residual hinge and the tail domain are from HsKin (NcKin_384$_{hTail}$).

**pNK433_340-79_hktail:** PCR on pHK546_N$_2$K$_2$ with HK560(KpnIΔSacI)- and HK560 (SacIΔPstI)+. The primer removed a *Sac*I-site at 772 and a *Pst*I-site at 856 in the human kinesin sequence and introduced a *Kpn*I-site and a *Sac*I-site, C- and N-terminal to the neck domain, respectively. Cloning via *Kpn*I and *Sac*I into pNK433_C307A_hktail. The resulting construct contains the NcKin head, neck-linker and hinge domain and the HsKin neck domain (NcKin_340-79$_{hTail}$).

**pNK433_340_hktail:** PCR on pHK546_ N$_2$K$_2$ with HK560(Pst)- and HK560(Kpn)+. Cloning via *Pst*I and *Kpn*I into pNK433_340-79_hktail, thereby introducing the human kinesin hinge. The resulting construct contains the NcKin head and neck-linker and the HsKin neck and hinge domain (NcKin_340$_{hTail}$).

All clones used for protein expressions were verified by sequencing.

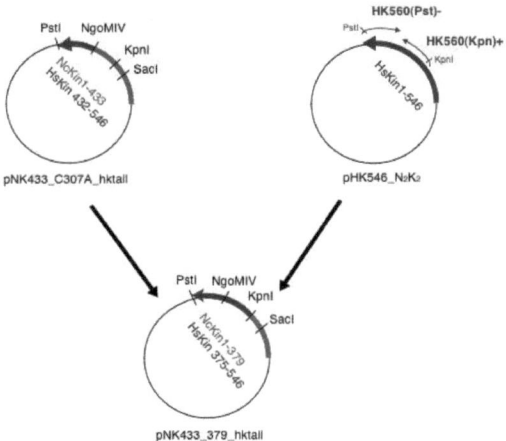

**Fig. 2.1: Cloning of pNK433_379_hktail.**
HsKin sequences were amplified via PCR on the pHK546_N$_2$K$_2$ plasmid using HK560(PstI)- and HK560(Kpn)+ as primer. The fragment was cleaved with PstI and KpnI and ligated into the parent vector pNK433-C307A_hktail that had been cut with the same enzymes. The resulting plasmid contains NcKin sequences from 1 – 379 (head, neck-linker and neck domains) and HsKin sequences from 375 to 546 (hinge and tail domains).

### 2.2.12.4 Generation of pNcKin_stableNeck

In the NcKin_stableNeck construct, the NcKin neck domain between aa 342 and 374 was replaced by 33 aa of an artificial coiled-coil sequence, called the EIEALKA sequence. The DNA fragment for this artificial sequence was generated via two synthetic oligonucleotides, coiled oligo- and coiled oligo+ (Fig. 2.2). Coiled oligo- represents the anti-sense strand and encodes heptads III to V of the artificial neck. Coiled oligo+ represents the sense-strand and encodes heptads I to III. In heptad III, the two oligonucleotides overlap in a complementary sequence. To ensure specific annealing, different codons for the residues EIEALKA were chosen in heptad III than in the other heptads. After annealing of the synthetic oligonucleotides, the single-stranded parts were filled up by DNA-polymerase (2.2.10), resulting in a double-stranded DNA fragment with 99 bp, encoding for 5 heptads of the artificial coiled-coil sequence. Before cloning into the vector, a *Kpn*I-site C-terminal (3'end) and a *Sac*I-site N-terminal (5'end) to the neck sequence had to be introduced. This was done via PCR on the synthetic DNA fragment using stablecoil+ and stablecoil- as primer. The amplified fragment (133 bp) was then cleaved with *Kpn*I and *Sac*I and ligated into the pNK433_C307A_hktail vector that had been cut before using the same enzymes.

Clones used for protein expressions were verified by sequencing.

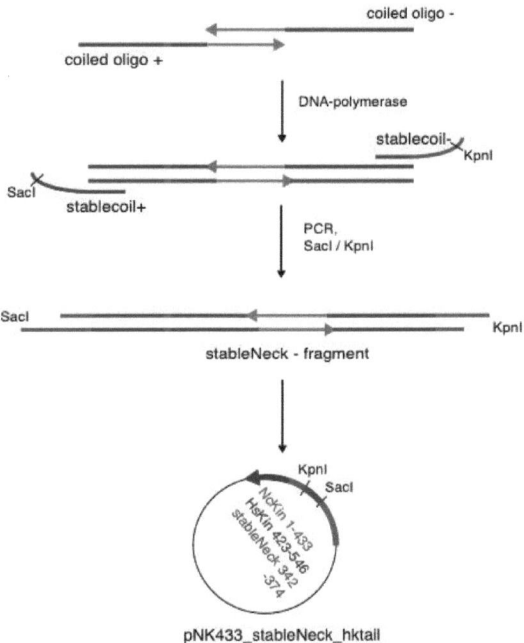

**Fig. 2.2: Cloning of pNK433_stableNeck_hktail.**
The artificial neck sequence was generated via the synthetic oligonucleotides coiled oligo+ / - that were annealed and filled up with DNA-ploymerase. The *Kpn*I and *Sac*I sites were introduced via PCR with stablecoil+ and stable coil- as primer. The resulting fragment was ligated into the parent vector pNK433_C307A_hktail.

### 2.2.12.5 Generation of chimeric constructs as 433-versions

All chimeric constructs including NcKin_stableNeck were cloned as two versions: one with a part of the human tail domain appending C-terminal to the protein (NcKin$_{hTail}$), which readily adsorb on glass surfaces, and one ending at aa 433 with a C-terminal cys-tag (NcKin$_{433}$), used for determination of the oligomerization state. The constructs were originally cloned into the hTail background, as described in 2.2.12.3 – 4. To generate the 433-versions, the hTail domain was removed using *Pst*I and *Ngo*MIV. The cys-tag (PSIVHRKCF, 2.2.12.1) was then introduced via synthetic oligonucleotides, cystagNgoMIV- and cystagNgoMIV+. The oligonucleotides were complementary to each other and *Ngo*MIV -compatible at the 5'end and *Pst*I compatible at the 3'end (referring to the sense-strand). After annealing, the double-stranded fragment was cleaved and ligated into the plasmids encoding the chimeric constructs, where the hTail domain had been removed before via *Pst*I and *Ngo*MIV.

All clones used for protein expressions were verified by sequencing.

### 2.2.12.6 Generation of P342C-constructs

All P342C-constructs are based on the plasmid pNK433_C307A_P342C_hktail, constructed by K. Hahlen [Hahlen, 2004a]. A fragment containing the P342C mutation was cleaved out of this plasmid via *Bam*HI and *Kpn*I, and cloned into the plasmids pNK433_379_hktail and pNK433_384_hktail, yielding **pNK433_379_P342C_hktail** and **pNk433_384_P342C_ hktail**. The double mutant **pNK433_Y362K,P342C_hktail** was also provided by Katrin Hahlen.

For generation of the 433-versions **pNK433_379_P342C and pNK433_384_P342C**, the constructs were subcloned via *Ngo*MIV and *Bam*HI into the plasmid pNK433_C307A.

### 2.2.12.7 Summary of all constructs used in this work

| Plasmid Name | Protein Name | Comment |
| --- | --- | --- |
| pNK433cys | NcKin$_{433}$ | NcKin wild-type, 1-433; cys-tag; [Kallipolitou, 2002] |
| pNK433_C307A_cys | NcKin$_{433}$ | NcKin 1-433 with C307A; cys-tag |
| pNK433_C307A_hktail | NcKin$_{hTail}$ | NcKin 1-433 with C307A; hTail domain; [Hahlen, 2004a] |
| pNK433cys-Y362K | NcKin$_{433}$Y362K | NcKin 1-433 with Y362K; cys-tag (G. Woehlke) |
| pNK433-Y362C | NcKin$_{433}$Y362C | NcKin 1-433 with Y362C |
| pNK433cys-Y362F | NcKin$_{433}$Y362F | NcKin 1-433 with Y362F; cys-tag |
| pNK433-F359C | NcKin$_{433}$F359C | NcKin 1-433 with F359C |

| | | |
|---|---|---|
| pNK344his | NcKin$_{344}$ | NcKin wild-type, 1-344; his-tag |
| pHK339his | HsKin$_{339}$ | HsKin wild-type, 1-339; his-tag |
| pNK433_340_hktail | NcKin_340$_{hTail}$ | NcKin 1-340; HsKin 336-546, including hTail |
| pNK433_379_hktail | NcKin_379$_{hTail}$ | NcKin 1-379; HsKin 375-546, including hTail |
| pNK433_384_hktail | NcKin_384$_{hTail}$ | NcKin 1-384; HsKin 380-546, including hTail |
| pNK433_340-79_hktail | NcKin_340-79$_{hTail}$ | NcKin 1-340 and 380-433; HsKin 336-374, including hTail |
| pNcKin433_stableNeck_hktail | NcKin_stableNeck$_{hTail}$ | NcKin 1-433 with artificial neck sequence 342-374; hTail |
| pNK433cys_340 | NcKin_340$_{433}$ | NcKin 1-340 and 424-433; HsKin 336-417; cys-tag |
| pNK433cys_379 | NcKin_379$_{433}$ | NcKin 1-379 and 424-433; HsKin 375-417; cys-tag |
| pNK433cys_384 | NcKin_384$_{433}$ | NcKin 1-384 and 424-433; HsKin 380-417; cys-tag |
| pNK433cys_340-79 | NcKin_340-79$_{433}$ | NcKin 1-340 and 380-433; HsKin 336-374; cys-tag |
| pNK433cys_stableNeck | NcKin_stableNeck$_{433}$ | NcKin 1-433 with artificial neck sequence 342-374; cys-tag |
| pNK433_C307A_P342C_hktail | NcKin$_{hTail}$P342C | NcKin 1-433 with C307A and P342C; hTail; [Hahlen, 2004a] |
| pNK433_Y362K,P342C_hktail | NcKin$_{hTail}$ Y362K,P342C | NcKin 1-433 with C307A, Y362K and P342C; hTail; [Hahlen, 2004a] |

| | | |
|---|---|---|
| pNK433_379_P342C_hktail | NcKin_380$_{hTail}$P342C | NcKin 1-379 with P342C, HsKin 375-546, including hTail |
| pNK433_384_P342C_hktail | NcKin_384$_{hTail}$P342C | NcKin 1-384 with P342C, HsKin 380-546, including hTail |
| pNK433_379_P342C | NcKin_380$_{433}$P342C | NcKin 1-379 and 424-433, with P342C, HsKin 375-417 |
| pNK433_384_P342C | NcKin_384$_{433}$P342C | NcKin 1-384 and 424-433, with P342C, HsKin 380-417 |

## 2.3 Biochemical methods

### 2.3.1 SDS-Polyacrylamide gel electrophoresis (SDS-PAGE)

Proteins were separated on discontinuous SDS-polyacrylamide gels containing 10 % polyacrylamide (PAA) [Laemmli, 1970]. The gels were run in the "Multigel-Long-System" (Biometra, München) at 50 – 100V. Samples and high molecular weight standard (Sigma) were mixed with Laemmli sample buffer [Laemmli, 1970], incubated at 95°C for 5 min and immediately loaded onto the gel.

PAA solution:                30% acrylamide, 0.8% bisacrylamide (Biorad)
Running buffer:              25 mM Tris•HCl, 0.1% SDS, 192 mM glycine
10x buffer for stacking gel: 500 mM Tris•HCl, pH 6.8, 0.4% SDS
10x buffer for separating gel: 1.5 M Tris•HCl, pH 8.8, 0.4% SDS
6x Laemmli sample buffer:    300 mM Tris•HCl, pH 6.8, 15 mM EDTA, 12% SDS, 30% glycerol,
                             15% β-mercaptoethanol, 0.06% bromphenol blue

Non-reducing SDS-PAGE was carried out to prove inter- and intra-molecular disulfide bridges in NcKin constructs. These crosslinks are covalent bonds, which are not disrupted by heat but by reducing agents. Therefore, a 2x Laemmli sample buffer without β-mercaptoethanol was used. Band shifts between reduced and oxidized samples indicate different conformations of the same construct.

### 2.3.2 Staining of SDS gels

#### 2.3.2.1 Coomassie staining

Gels were stained for 30-60 min in Coomassie staining solution, rinsed with $H_2O$ and destained with 10% acetic acid. For documentation gels were photographed with a CCD camera (Eagle Eye System, Stratagene) or scanned (Epson 1200 Photo).

Coomassie staining solution: 7.5% acetic acid, 50% methanol, 0.25% Coomassie Brilliant Blue
                             R250 (Sigma)

### 2.3.2.2 Colloidal Coomassie staining

SDS-gels for sucrose-density centrifugations were stained with the "Colloidal Blue Staining Kit" (Invitrogen, Paisley) following the instructions of the manufacturer. After staining for at least 12 h, the gels were de-stained in clear water for 24 h. Gels were photographed with a CCD camera (Eagle Eye System, Stratagene) or scanned (Epson 1200 Photo) prior to analysis.

## 2.3.3 Expression of kinesin constructs

Media flasks with 2 l TPM were inoculated from a freshly transformed single colony of *E. coli* BL21 CodonPlus (DE3)-RIL (Stratagene Inc.). Cells were pre-grown for 16-20 h at 22°C until an $OD_{600}$ of 0.4-0.8 for NcKin constructs and 1.0-1.5 for HsKin constructs was reached. Expression was induced with 1 mM isopropyl-β-D-thiogalactopyranoside (IPTG) and cell cultures incubated overnight at 22°C. The next morning cells were harvested (3500 rpm, 35 min, 4°C) and stored at – 70°C.

## 2.3.4 Protein purification

### 2.3.4.1 Chromatographic purification of bacterially expressed NcKin constructs

NcKin proteins were purified by ion exchange chromatography on a 5 ml HiTrap SP-Sepharose column on a FPLC system (both Amersham Pharmacia Biotech) according to a protocol of [Crevel et al., 1999]. 3-5 g frozen cells were resuspended in 3 fold volume lysis buffer. The suspension was sonified (4x 30 s; output: 4; duty cycle: constant, Branson Sonifier 250) and spun clear of debris (Beckmann rotor 42.1: 42000 rpm, 45 min, 4°C).

The FPLC column was first washed with $H_2O$ and then equilibrated with buffer A. For calibration, the conductivity meter was set to 100% ionic strength with buffer B and to 0% ionic strength with buffer A.

The supernatant was diluted 3.5fold with $H_2O$ to 20 mM salt with additional 5 mM $MgCl_2$, 1 mM DTT, 0.5 mM EGTA, 10 μM ATP. This mixture was immediately loaded onto the column using a peristaltic pump (Amersham Pharmacia Biotech, 4 ml/min). Unbound protein was removed by washing the column with buffer A. The kinesin was eluted in a manual step gradient from 50 mM to 1 M NaCl and collected in 1.5 ml fractions. Peak fractions of active kinesin were identified by SDS-PAGE (2.3.1) and ATPase assay (XXX). Monomeric NcKin constructs typically eluted with 100 mM, dimeric constructs with 200 mM NaCl. The peak fractions were pooled, supplemented with 1/10 volume of glycerol, frozen in liquid nitrogen in small aliquots and stored at -70°C.

Lysis buffer: buffer A with 50 mM NaCl, 0.5 $\mu$M Pefabloc, 1x protease inhibitor (PI), 1 mM DTT, 0.5 mM EGTA, lysozyme and DNAse I
100x PI: 1 mg/ml soybean trypsine inhibitor, 1 mg/ml TAME, 250 $\mu$g/ml leupeptine, 100 $\mu$g/ml pepstatine A, 100 $\mu$g/ml aprotinine
Buffer A: 20 mM Na-Phosphat, 5 mM $MgCl_2$, 10 $\mu$M ATP, pH 7.4
Buffer B: buffer A with 1 M NaCl

Buffers A and B were filtered to remove gas and particles.

### 2.3.4.2 Chromatographic purification of bacterially expressed, his-tagged HsKin constructs

Purification of HsKin constructs was carried out in a two step chromatographic process. First, the protein was accumulated via its histidin tag on a Ni-NTA (Qiagen) column. Second, the eluat was further purified with a 5 ml HiTrap Q-Sepharose column on a FPLC system (both Amersham Pharmacia Biotech).

5-10 g cells were resuspended in the 4 fold volume of lysis buffer. The suspension was sonified (4x 30 s; output: 4; duty cycle: constant, Branson Sonifier 250) and spun clear of debris (Beckmann rotor 42.1: 42000 rpm, 45 min, 4°C).

In the meantime, 6 ml of the Ni-NTA material was sedimented (Heraeus Minifuge RF: 500 rpm, 2 min, 4°C) and washed 2 times in 15 ml lysis buffer without ATP and imidazole. Then the protein was bound to the resin for 45-60 min at 4°C. The mixture was loaded onto a Qiagen disposable column and washed with 50 ml Ni wash buffer until the eluate displayed an $E_{280}$ less than 0.1. The protein was eluted with 10x 1 ml fractions of Ni elution buffer. The fractions with the highest $E_{280}$ were pooled and diluted 4 fold with MonoQ buffer A + 1 mM DTT.

The FPLC was prepared as in 2.3.4.1 and equilibrated with 5% MonoQ buffer B (50 mM ionic strength). The diluted Ni-NTA eluate was immediately loaded onto the column using a peristaltic pump (Amersham Pharmacia Biotech, 4 ml/min). Unbound protein was removed by washing the column with buffer A. The kinesin was eluted in a manual step gradient from 50 mM to 1 M NaCl and collected in 1.5 ml fractions. The peak fractions of active kinesin were identified by an ATPase assay (2.3.13.2) and their purity checked by SDS-PAGE (2.3.1) Typically, monomeric HsKin constructs eluted between 100 and 200 mM. The peak fractions were pooled, supplemented with 1/10 volume of glycerol, frozen in liquid nitrogen in small aliquots and stored at -70°C.

| | |
|---|---|
| Lysis buffer: | 50 mM Na-Phosphat, 250 mM NaCl, 2 mM $MgCl_2$, 1 mM ATP, 20 mM imidazole, 10 mM β-mercaptoethanol, 0.5 μM Pefabloc, 1x PI, pH 8.0 |
| Ni wash buffer: | 50 mM Na-Phosphat, 250 mM NaCl, 1 mM $MgCl_2$, 0.1 mM ATP, 10 mM β-mercaptoethanol, pH 6.0 |
| Ni elution buffer: | 50 mM Na-Phosphat, 250 mM NaCl, 1 mM $MgCl_2$, 0.1 mM ATP, 500 mM imidazole, 10 mM β-mercaptoethanol, pH 7.2 |
| MonoQ buffer A: | 25 mM PIPES·KOH, 2 mM $MgCl_2$, 1 mM EGTA, 0.1 mM ATP, pH 6.8, |
| MonoQ buffer B: | MonoQ buffer A with 1 M NaCl |

### 2.3.4.3 Affinity purification of kinesin constructs

To select for active motor proteins, kinesin constructs used in single molecule assays were purified via a microtubule binding and release step. 3-5 g cells were resuspended in the 3 fold volume of AP100+, sonified (4x 30 s; output: 4; duty cycle: constant, Branson Sonifier 250) and spun to remove cell debris (Beckmann rotor 42.1: 42000 rpm, 45 min, 4°C). 20 μM taxol (paclitaxel, Sigma T-7402), taxol-stabilized microtubules (0.3–0.5 mg/ml), 200 μM AMP-PNP and apyrase (5 U/ml) were added to the supernatant and incubated for 15–30 min at room temperature. The kinesin-microtubule-complexes were sedimented (Beckmann rotor 42.1: 42000 rpm, 30 min, 22°C) and the supernatant discarded. The pellet was washed once with AP100, resuspended in 1 ml KCl-buffer and spun through a sucrose cushion to remove unpolymerized tubulin (Optima rotor 100.3: 80000 rpm, 10 min, 22°C). The pellet was resuspended in 200 μl release buffer and incubated 15 min at room temperature to ensure complete release of active motors. Microtubules were sedimented (Optima rotor 100.3: 80000 rpm, 10 min, 22°C) and the supernatant containing the active motor proteins was removed and stored on ice. The release procedure was repeated once. Kinesin constructs in the release-fractions were identified via SDS-PAGE and – in case of similar concentrations in both fractions – pooled. Aliquots were supplemented with 10 % glycerol, frozen in liquid nitrogen and stored at –70°C.

| | |
|---|---|
| AP100: | 100 mM PIPES•KOH, 2 mM $MgCl_2$, 1 mM EGTA, pH 6.8 |
| AP100+: | AP100 with 1 mM DTT, 200 μM Pefabloc (Roche Diagnostics), 1x PI |
| Sucrose cushion. | 40 % sucrose in AP100 |
| KCl-buffer: | AP100 with 50 mM KCl, 10 μM taxol, 200 μM AMP-PNP |
| Release-buffer: | AP100 with 10 mM ATP, 10 mM $MgCl_2$, 200 mM KCl, 10 μM taxol |

### 2.3.4.4 Purification of pig brain tubulin

Pig brain tubulin was purified in three successive steps of polymerisation and depolymerization, followed by ion exchange chromatography [Mandelkow et al., 1985]. Fresh pig brains were obtained at the local slaughterhouse, immediately put on ice and separated from blood vessels and connective tissue. A mixture of 700 g of brain with 700 ml buffer A was homogenised in a warring blender (Braun) and centrifuged (GSA-rotor: 13000 rpm, 4°C, 70 min). The supernatant was inoculated with 25% glycerol (final concentration v/v, Roth 3783.1) and 2 mM ATP (final concentration). To polymerize the tubulin, this mixture was incubated for 30 min by slightly shaking at 35°C. The microtubules were sedimented at 32°C (pre-warmed centrifuges and Beckmann rotors 35: 35000 rpm, 50 min and TI 45: 42000 rpm, 45 min). The pellets were resuspended in 100 ml buffer C and homogenised on ice in dounce homogenisers (Wheaton). The microtubules were allowed to depolymerize on ice for 25 min and centrifuged again (pre-cooled centrifuge and Beckmann rotor 42.1: 36000 rpm, 30 min, 4°C). After addition of 2 mM ATP the supernatant was polymerized once more at 35°C for 30 min and pelleted again (pre-warmed centrifuge and Beckmann rotor 42.1: 33000 rpm, 60 min, 32°C). The pelleted microtubules were weighed, frozen in liquid nitrogen and stored at -70°C.

To remove last traces of microtubule-binding proteins, the tubulin was passed through a phosphocellulose column. 50 ml of the activated phosphocellulose (P-11, Whatman) material (150 ml total volume) was packed into a column (Amersham-Pharmacia) and equilibrated with 3 volumes of buffer D at 1 ml/min in the FPLC (Amersham-Pharmacia). In the meantime microtubules were thawed and homogenised on ice in 50-100 ml buffer B. For microtubule depolymerization the mixture was incubated on ice for 25 min. After centrifugation (pre-cooled centrifuge and Beckmann rotor 42.1: 36000 rpm, 30 min, 4°C), the polymerization of the supernatant was induced with 10% DMSO and 2 mM ATP and the mixture was slightly shaken at 35°C for 30 min. The microtubules were pelleted (pre-warmed centrifuge and Beckmann rotor 42.1: 33000 rpm, 60 min, 32°C) and resuspended in 5-7 ml buffer D, again homogenised and depolymerized on ice for 25 min. After centrifugation (pre-cooled centrifuge and Beckmann rotor TI 70.1: 34200 rpm, 30 min, 4°C) the supernatant was loaded onto the column using a peristaltic pump (Amersham-Pharmacia). After the sample had been loaded completely, the column was washed with buffer D with 0.16 ml/min. In contrast to microtubule associated proteins tubulin does not bind to phosphocellulose under these buffer conditions. 1 ml fractions were collected and peak fractions were identified using the Bradford reagent (2 $\mu$l sample with 400 $\mu$l $H_2O$ and 100 $\mu$l Bradford reagent. Peak fractions were pooled, supplemented with 0.1 mM GTP and frozen in liquid nitrogen. The tubulin was stored at -70°C.

| | |
|---|---|
| Buffer A: | 0.1 M PIPES·NaOH, 2 mM EGTA, 1 mM MgSO$_4$, 1 mM DTT, 100 µM ATP |
| Buffer B: | 0.5 M PIPES·NaOH, 1 mM EGTA, 1 mM MgSO$_4$, 1 mM DTT, 1 mM ATP |
| Buffer C: | 0.1 M PIPES·NaOH, 1 mM EGTA, 1 mM MgSO$_4$, 1 mM DTT, 1 mM ATP |
| Buffer D: | 0.1 M PIPES·NaOH, 1 mM EGTA, 1 mM MgSO$_4$, 1 mM DTT, 50 µM ATP |

All buffers were adjusted to pH 6.9 at 4°C.

### 2.3.5  Determination of protein concentration

The protein concentration was determined with Bradford reagent (Biorad, 500-0006) [Bradford, 1976]. For each measurement, a standard curve with bovine serum albumine (BSA) was measured in parallel. The absorption at 630 nm was determined in a microplate reader (Dynatech MR 5000) and the protein concentration was calculated using the BSA standard as reference. Kinesin concentrations are given in monomer concentrations.

### 2.3.6  Determination of the oligomerization state

To characterize the oligomerization state of NcKin constructs, the molecular weights were estimated by gel filtration and sucrose density centrifugation.

#### 2.3.6.1  Sucrose density centrifugation

Sedimentation coefficients $S_{w,20}$ were measured by sucrose density centrifugation. The gradient was generated by layering 500 µl of each 18, 15, 11, 9, 7 and 5% sucrose solutions in buffer A, starting with the highest sucrose concentration (total gradient volume: 3500 µl). Protein solutions (300 µl) of app. 5 µM concentration were loaded on the sucrose gradient and centrifuged for 16 h (Beckman rotor SW 50.1: 37000 rpm; 4°C). As internal standards aldolase, BSA and carboanhydrase ($S_{w,20}$ = 7.4, 4.3 and 3.2, Roche Diagnostics) were included. The gradient was fractionated using a peristaltic pump (Amersham-Pharmacia) and analysed by SDS-PAGE (2.3.1). The position of the protein peaks was identified by densitometrical analysis using the NIH-Image software. $S_{w,20}$ values of kinesin samples were determined with the standard proteins as reference (Kaleidagraph software).

| | |
|---|---|
| Buffer A: | 20 mM Na-Phosphat, 200 mM NaCl, 5 mM MgCl$_2$, 10 µM ATP, 1 mM DTT, pH 7.4 |

### 2.3.6.2 Gel filtration

Stokes radii ($r_{stokes}$) were determined by gel filtration analysis. Protein solutions (500 µl, 8-15 µM) were loaded on a Sephadex 200 column (Amersham Pharmacia), equilibrated with buffer A (2.3.8.1, without DTT). Elution volumes of kinesin samples were compared with standard proteins (ferritin: 5.9 nm $r_{stokes}$, aldolase: 4.5 nm, BSA: 3.5 nm, carboanhydrase: 2.4 nm, cytochrome c: 1.6 nm; Roche Diadnostics and Sigma). Stokes radii were determined from a plot of elution volumes versus standard sizes [Andrews, 1970]. The molecular weight was then calculated according to the equation of [Cantor et al., 1980]:

$$M_r = (S_{w,20} \times n_a \times 6\pi \times \eta \times r_{stokes})/(1 - \upsilon \times \rho)$$

with $n_a$ = Avogadro number ($6.023 \cdot 10^{23}$ mol$^{-1}$), $\eta$ = viscosity ($H_2O$ at 293 K: $10^{-3}$ N·s·m$^{-2}$), $\upsilon$ = specific volume of the protein ($0.725$ cm$^3$·g$^{-1}$) and $\rho$ = density of the medium ($H_2O$ at 293 K: 1 g·cm$^{-3}$)

Insertion of these numbers yields:

$$M = 4128 \, S_{w,20} \cdot r_{stokes} \, [g/mol]$$

### 2.3.7 Polymerization of microtubules

For accurate determination of the microtubule concentration, polymerization-inactive tubulin had to be removed. Aggregated tubulin was pelleted in a cold spin (pre-cooled centrifuge and Beckmann rotor TLA 100.3: 80000 rpm, 4°C, 10 min). The supernatant was polymerized with 1 mM GTP at 37°C for 10 min. Then, 20 µM taxol (paclitaxel, Molecular Probes) was added for stabilization of the microtubules and incubated for another 20 min at 37°C. To remove unpolymerized tubulin, the microtubules were sedimented through a 40% sucrose cushion (pre-warmed centrifuge and Beckmann rotor TLA 120.1: 80000 rpm, 25°C, 10 min). The microtubule pellet was washed once with the assay buffer containing 20 µM taxol and then resuspended in this buffer. The volume was adjusted to achieve a final tubulin concentration of app. 100 µM tubulin dimer.

In some cases the accurate microtubule concentration was not relevant for the assay. Here, tubulin was thawed and quickly polymerized by adding 1 mM GTP and incubated 25 min at 37°C. For microtubule stabilization 20 µM taxol was added after 10 min.

## 2.3.8 Determination of microtubule concentration

Microtubule concentrations were determined in 6 M guanidium·HCl by measuring absorption at 280 nm in the photometer.

Microtubules were diluted 1:10 and 1:5 in assay buffer containing 20 $\mu$M taxol, 10 $\mu$l of each dilution was mixed with 90 $\mu$l of 6.6 M guanidium·HCl and then measured in the photometer.

Microtubule concentration was calculated as follows [Huang et al., 1994a]:

$(E_{280} / 10.3) \cdot$ dilution $= \mu$M MT

## 2.3.9 Fluorescent labelling of tubulin

To visualize microtubules in the fluorescence microscope (2.3.12), purified pig brain tubulin was labelled with the fluorescent dye Cy5. The labelling procedure involved coupling the reactive Cy5-succinimidyl ester (Cy5-OSu monofunctional reactive fluorophore, Amersham Pharmacia, PA13600) to polymerized microtubules, thereby protecting residues important for microtubule assembly (revised from T. Mitchison lab protocols, personal communication). The labelling was performed at high pH to optimize the reaction with succinimidyl esters and functional tubulin was selected after the labelling reaction by several cycles of polymerization and depolymerization. Typically, the yield of labelled tubulin was very low (1-3 %). Thus, the protocol had to be started with sufficient amounts of tubulin. It was very critical to strictly keep the different temperatures during polymerization and depolymerization steps.

3-5 ml of purified tubulin (app. 80 mg) were thawed quickly, supplemented with GTP to 1 mM and stored on ice for 5 min. To promote polymerization, DMSO to a final of 10% (v/v) was added and the mixture transferred to 37°C for 30 min. The polymerized tubulin was layered onto 4 ml high-pH cushion in two 100.3 Ti tubes (Beckmann). Microtubules were sedimented in a pre-warmed centrifuge (Optima rotor 100.3, 80000 rpm, 15 min, 25°C). During the centrifugation, the Cy5-OSu was dissolved in anhydrous DMSO (Amersham 5-pak in a total of 50 $\mu$l). After the spin, the pH cushion was carefully removed, the pellet washed with warm BRB80+/taxol and then resuspended in 1-2 ml labelling buffer using the pistil of a small homogeniser (Wheaton). At this stage it was very critical to keep the temperature at 37°C (resuspension was performed in a water bath). For labelling, a 10-20 fold molar excess of Cy5-OSu was added (5-pack for app. 25 mg tubulin) and incubated for 30-40 min at 37°C, while gently mixing every 2-3 min. At end of the incubation an equal volume of quench-solution was added to the reaction, mixed well and incubated for another 5 min. The quenched labelling reaction was then layered onto 3 ml of low pH cushion in two Ti 100.3 tubes and spun in a pre-warmed centrifuge (Optima 100.3 rotor, 80000 rpm, 10 min, 25 °C). After the spin, the cushion was carefully removed, the pellet washed with warm BRB80+/taxol and resuspended in 1 ml of ice-cold 1 x IB-buffer using a pre-cooled dounce homogeniser (2-5 ml volume, Wheaton) in an ice-water bath. Douncing was performed until the suspension was uniform and continued for 30 min to allow for depolymerization. The tubulin was then spun in a pre-cooled centrifuge (Optima 100.3 rotor, 80000 rpm, 10 min, 2°C). The supernatant was removed carefully and supplemented to 1 x BRB80+ (from a 5 x stock) and 1 mM GTP and incubated on ice for 3 min. The mixture was then warmed for 2 min at 37°C, glycerol was added to 33% final (v/v, Roth 3783.1) and microtubules were polymerized for 30 min at 37°C. The polymerization reaction was layered onto 1 ml low pH cushion and spun (pre-warmed Optima rotor 100.3: 80000, 15 min, 25°C). After the centrifugation, the cushion was removed carefully, the pellet washed with 1 ml warm IB-buffer, resuspended in 0.2-0.4 µl ice-cold IB and incubated for 20-30 min on ice. The depolymerized tubulin was spun (pre-cooled Optima rotor 100.3: 80000 rpm, 10 min, 2°C). The supernatant was recovered and frozen in liquid nitrogen in small aliquots (3-5 µl). Labelled tubulin was stored at −70°C in a box shielded from light.

| | |
|---|---|
| High pH cushion: | 0.1 M NaHEPES, 1mM $MgCl_2$, 1 mM EGTA, 60% glycerol, pH 8.6 |
| Labelling buffer: | 0.1 M NaHEPES, 1mM $MgCl_2$, 1 mM EGTA, 40% glycerol, pH 8.6 |
| 5 x BRB80+: | 400 mM PIPES, 25 mM $MgCl_2$, 5 mM EGTA, pH 6.8 |
| Quench: | 2 x BRB80+, 100 mM K-Glutamate, 40% glycerol |
| Low pH cushion: | 1x BRB80+ with 60% glycerol |
| 10 x IB-buffer: | 500 mM K-Glutamate, 5 mM $MgCl_2$, pH 7.0 |

### 2.3.9.1 Estimation of yield and labelling stoichiometry

To estimate the tubulin concentration and stoichiometry of labelling, the Cy5-tubulin was diluted 1:100 in IB-buffer and the absorption was measured at 280 nm (protein excitation) and 649 nm (Cy5 excitation). Using the extinction coefficients for tubulin dimers (10.3 /µM·cm, 2.3.8) and Cy5 (250000 /cm·M, provided by the manufacturer) the molar concentrations of tubulin and fluorescent dye could be calculated. The typical yield of tubulin was 300-400 µM in a total volume of 200-500 µl. The labelling ratio was typically 30-45%.

## 2.3.10 Biotinylation of cys-tagged NcKin constructs

The cysteine-tagged NcKin$_{433}$ constructs were labelled with biotin-maleimide [Kallipolitou et al., 2001] to perform gliding assays on a streptavidin-coated coverslip (2.3.4.2). The protein was incubated with a 6 fold molar excess of maleimide conjugate on ice for 60 min. The reaction was stopped with 10 mM DTT. Active kinesin was isolated by a microtubule binding and release step [Vale et al., 1985], similar to the affinity purification protocol (2.3.4.3). Microtubules were polymerized as described abbove and resuspended in BRB80+ buffer containing 20 $\mu$M taxol. The kinesin constructs were incubated with a 3-4 fold molar excess of microtubules and 2 mM AMP-PNP for 10 min at room temperature. Microtubule-kinesin complexes were sedimented (Optima rotor 100.3: 80000 rpm, 10 min 25 °C) through a 40 % sucrose cushion in BRB80. To release active motor proteins, the pellet was washed once in BRB80+/taxol and then resuspended in 100 $\mu$l release buffer. Microtubules were sedimented again (Optima rotor 100.3: 80000 rpm, 10 min 25 °C) to separate released motors and microtubules. The release procedure was repeated once. Kinesin constructs in the release-fractions were identified via SDS-PAGE and – in case of similar concentrations in both fractions – pooled. Aliquots were supplemented with 10 % glycerol, frozen in liquid nitrogen and stored at –70°C.

BRB80+: 80 mM PIPES·KOH, 1 mM MgCl$_2$, 1 mM EGTA, pH 6.8
Release buffer: BRB80+ with 200 mM KCl, 20 $\mu$M taxol, 10 mM ATP, 10 mM MgCl$_2$

## 2.3.11 Multiple motor gliding assay

### 2.3.11.1 Video enhanced light microscopy

Light microscopy was carried out in a temperature-controlled Zeiss Axiophot microscope with a 63x oil immersion objective (Zeiss, Oberkochen). Usually, gliding assays were performed at 25 °C. Microtubules were observed in phase contrast with a Newicon C2400-7 camera (Hamamatsu). The image was transferred via a "DVS 1000 Image Processing System" (Hamamatsu) to a monitor (Sony) and recorded on videotape (Fuji). The "DVS 1000 Image Processing System" was used for contrast enhancement and background subtraction. Gliding velocities were determined by measuring the time a microtubule needed for movement between two marked points on the monitor. Distances on the monitor were calibrated using an object micrometer.

### 2.3.11.2 Gliding assay with biotin-labelled NcKin constructs

The velocity of kinesin constructs was measured in a microscopic motility assay [Paschal et al., 1993]. For biotin-labelled kinesins the assay was performed in a "flow cell" coated with streptavidin. Flow cells consist of an object holder with a coverslip stuck via two thin stripes of silicone grease or "double-stick" tape, forming a chamber of 5-15 $\mu$l volume that was open at two ends. That way the sample could be floated into the chamber and proteins that were not adsorbed onto the glass surface could be washed out. To prevent evaporation of the sample, the flow chamber was closed by wax or two more stripes of silicone grease.

Flow cells were incubated with 1 mg/ml streptavidin (Sigma Chemical Corp.) in BRB80+ for 5 min and then washed with three chamber volumes of blocking buffer. Biotin-labelled kinesin in a concentration of 20-30 $\mu$M in blocking was filled in and incubated for 5-10 min. The assay was started by floating the chamber with 3 volumes of motility buffer, containing the microtubules. Under these conditions, the motor density on the coverslip is very high (estimated: on average 28000 motor molecules/$\mu$m$^2$), leading to a situation where one microtubule (1-10 $\mu$m in length) is moved by thousands of motors at the same time; thus the assay is called a multiple motor gliding assay. To determine the gliding velocities, at least 20 microtubules were measured and averaged.

Blocking buffer: BRB80+ with 1 mg/ml BSA, 0.8 mg/ml casein, 20 $\mu$M taxol

Motility buffer: BRB80+ with 100 mM KCl, 5 mM MgCl$_2$, 10 mM ATP, 1-3 $\mu$l of microtubule dilution (1:20-1:50 in BRB80+/taxol; microtubules were quickly polymerized according to 2.3.6)

### 2.3.11.3 Gliding assay with hTail-constructs

Kinesin constructs containing a part of the human kinesin tail domain (NcKin$_{hTail}$) readily adsorbed on glass surfaces and thus could be used in the gliding assay without biotin-label. 1.25 $\mu$g kinesin was applied to a coverslip that was kept in a wet chamber for 1-2 min to allow the kinesin tail to bind to the glass surface. Casein buffer was added to a final volume of 5 $\mu$l. Then 1 $\mu$l BRB80+ with 100 mM KCl, ATP and MgCl$_2$ (final concentration: 10 mM) and 1-3 $\mu$l microtubules (diluted 1:20-1:50, quickly polymerized) were added. The coverslip was put onto an object holder and sealed with a wax-mixture. This slide was observed under a microscope, and moving microtubules were analysed as described above (2.3.10.2). In this assay, the motor density on the coverslip is similarly high as in gliding assays using biotin-labelled constructs (28000/$\mu$m$^2$, 2.3.10.2).

Casein buffer: BRB80+ with 0.1 M glucose, 2.5 mg/ml casein

## 2.3.12 Single molecule bead assay

Single molecule measurements of NcKin constructs were carried out in an optical laser trap apparatus in combination with a fluorescence TIRF microscope built by Anabel Clemen and Johann Jaud from the Technical University of Munich. Details of the experimental setup are described in 3.4.2 as far as necessary for understanding the experimental results.

To detect single molecule events, kinesin was unspecifically adsorbed in very low densities onto carboxylated latex beads that were 0.5 $\mu$m in diameter (Polysciences, Warrington, USA). Affinity-purified kinesin constructs (2.3.4.3) were appropriately diluted (1:400-1:4000, depending on the motor concentration) and mixed with equal volumes of beads (diluted 1:5 in BRB80+) and casein (5 mg/ml) as a blocking reagent. This mixture was incubated for at least 5 min and could be stored on ice for a maximum of 30 min. Immediately before the measurement, 1 $\mu$l of the bead-motor mix was further diluted in 400 $\mu$l BRB80+, containing 0.1 mg/ml casein. 4 $\mu$l of this dilution was then mixed with 12 $\mu$l motility buffer and floated into the prepared flow cell.

To generate fluorescently labelled microtubules, 5 $\mu$l of Cy5-labelled tubulin (2.3.9) was mixed with 100 $\mu$l of unlabelled tubulin. Aggregates were sedimented in a cold-spin and microtubules were polymerized in presence of 1 mM GTP for 30 min at 37°C. For stabilization, 20 $\mu$M taxol was added after 10 min (2.3.6). To immobilize the microtubules on the coverslip of the flow cell, the cell was first incubated with a polyclonal anti-tubulin antibody from sheep (1:10 in BRB80+) (Acris Antibodies, Hiddenhausen) for at least 10 min in a wet chamber. After washing the flow cell with three volumes of BRB80+, a 1:50 dilution of labelled microtubules in BRB80+/taxol was filled in and incubated for 1-3 min. Unbound microtubules were removed by floating the chamber with three volumes of BRB80+ containing 1 mg/ml BSA. Binding of the microtubules to the coverslip was confirmed in the TIRF microscope before the chamber was filled with the motor-bead solution. Using the laser trap as optical tweezers the beads with the adsorbed motor molecules were deposited on the fixed microtubules, so movement and force production of single molecules could be measured by detecting the bead position. Kinesin concentrations had to be adjusted in such a way that app. 1/3 of the beads in the cell showed movement along the microtubule, which leads to a probability of 6% that a single kinesin molecule is active on the bead [Block et al., 1990].

Motility buffer: BRB80+ with 100 mM KCl, 2 mM ATP, 1 mg/ml casein, 1% glucose, 6 $\mu$g/ml glucose-oxidase, 1 $\mu$g/ml catalase, pH 6.8

## 2.3.13 ATPase measurements

### 2.3.13.1 Basal ATPase measurements

The ATPase rates in the absence of microtubules were measured by following the phosphate release, using a Malachite Green reagent for colorimetric detection of phosphate [Geladopoulos *et al.*, 1991]. Free phosphate concentrations were determined by comparison with an appropriate $NaH_2PO_4$ standard that was tested in parallel.

The ATPase assays were started by adding NcKin at 0.3 and 0.6 $\mu M$ to 1 mM ATP in ATPase buffer (12A25+), alongside with a blank without kinesin. The reactions were stopped after 10 s, 20 s, 30 s, 1 min, 2 min, 5 min and 30 min by mixing 20 $\mu l$ of the reaction with 80 $\mu l$ of 0.3 M perchloric acid. 100 $\mu l$ of the Malachite Green reagent was added and incubated for exactly 15 min. The absorption at 640 nm was determined in a microplate reader (Dynatech MR 5000) and the phosphate concentration was calculated using the $NaH_2PO_4$ standard as reference. The amount of released phosphate was plotted versus the time. The basal ATPase rates were derived from the slopes of fits of the linear part of time traces, corrected for the blank (without NcKin).

| | |
|---|---|
| Malachite Green reagent: | 0.3 g/l Malachite Green (Sigma Chemicals Corp.); 2 g/l Na-molybdate; 0.5 g/l Triton X-100 in 0.7 M HCl, The solution was filtered and stored at 4°C. |
| ATPase buffer (12A25+): | 12 mM ACES·KOH, 25 mM K-acetate, 2 mM Mg-acetate, 3 mM $MgCl_2$, 0.5 mM EGTA, pH 6.8 |
| Dilution buffer for kinesin: (phosphate-free) | 12A25+, 100 mM NaCl, 1 g/ml BSA |

### 2.3.13.2 Microtubule-stimulated ATPase assay

The microtubule-stimulated steady-state ATP turnover was measured in a coupled enzymatic assay under saturating ATP conditions and with varying microtubule concentrations. In this assay, ATP turnover is coupled to the oxidation of NADH to $NAD^+$ by the enzymes lactate dehydrogenase (LDH) and pyruvate kinase (PK) (Fig. 2.3) [Huang *et al.*, 1994a]. PK requires phosphoenolpyruvate (PEP) as co-substrate. The rate of NADH oxidation can be observed in the photometer (Kontron Uvicon 930) by a decrease in extinction at 340 nm.

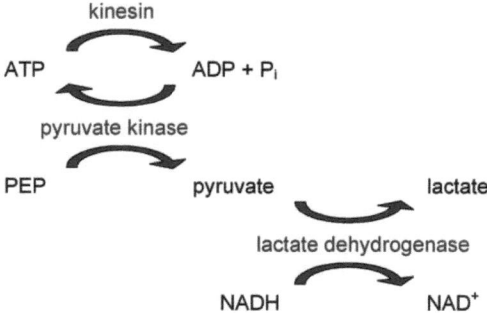

**Fig. 2.3: Priniciple of the enzyme coupled ATPase assay**
ATP hydrolysis by the kinesin is coupled to NADH oxidation via the reactions of lactate-dehydrogenase (LDH) and pyruvate-kinase (PK) in presence of non-limiting amounts of phosphoenolpyruvate (PEP). NADH oxidation was observed at 340 nm in the photometer.

A microtubule stock solution was prepared (2.3.7) and the concentration determined (2.3.9). The reaction was started by addition of 1 μl kinesin to 4 μl ATP, 4 μl NADH, 4 μl PEP, 2 μl enzyme mix, 0-20 μM microtubules (final concentration) and filled with 12A25+ to a whole assay volume of 80 μl. Kinesin was diluted 1:1 to 1:100, depending on the protein concentration, so that the enzymatic activities of LDH and PK were not rate-limiting in the assay. The reaction mixture was transferred quickly into a 50 μl quartz cuvette (Hellma) and the extinction $E_{340}$ was recorded for 1 min in the photometer (time drive mode). Its initial slope was determined (ΔE/min) and plotted against the corresponding microtubule concentration. Data points were fitted using a hyperbolic equation (Kaleidagraph software):

$\Delta E = \Delta E_{max} \cdot [Mt] / ([Mt]+K_{0.5,MT}) + B$

yielding the maximal rate of the extinction change ($\Delta E_{max}$ corresponding to the maximal reaction rate $V_{max}$) and the half maximal microtubule activation constant ($K_{0.5,MT}$). B stands for the background signal that was typically below 0.1% of the amplitude.

| | |
|---|---|
| 12A25+ buffer: | see 2.3.13.1 |
| ATP: | 20 mM ATP in 12A25+, pH 7.0 |
| NADH: | 4 mM nicotine adenine dinucleotide (Sigma) in 12A25+ |
| PEP: | 4 mg/ml phosphoenolpyruvate (mono potassium salt, Sigma) in 12A25+, pH 7.0 with 1 M KOH |

| Enzyme mix: | 30 µl lactate dehydrogenase (5.5 U/µl, glycerol solution, Roche), 30 µl pyruvate kinase (2 U/µl, glycerol solution, Roche), 90 µl 12A25+ |
|---|---|
| Dilution buffer for kinesin: | Buffer A (2.3.4.1) with 1 mg/ml BSA, 275 mM NaCl, 10 µM ATP |

### 2.3.13.3 Calculations for the coupled ATPase assay

*Calculation of $V_{max}$ from $\Delta E_{max}$/min*

Since NADH oxidation and ATP depletion are coupled 1:1 in the assay, the maximal ATP turnover per second could be derived from $\Delta E_{max}$ determined by the hyperbolic fit via the law of Lambert-Beer:

$$V_{max} = d \cdot \varepsilon \cdot \Delta E_{max} / 60 \text{ s}$$

$V_{max}$: maximal ATP turnover per second
$\Delta E_{max}$: maximal rate of NADH oxidation per minute
d: thickness of the cuvette (1 cm)
$\varepsilon$: extinction coefficient for NADH (6.22·mM$^{-1}$·cm$^{-1}$)

*Calculation of the maximal turnover rate ($k_{cat}$)*

The maximal turnover rate in the steady-state, $k_{cat}$, represents the maximal ATP turnover for a single kinesin head in one second and is therefore a characteristic value for each construct. It could be derived by dividing $V_{max}$ with the kinesin monomer concentration in the assay, using the protein concentration determined as described in 2.3.5.

$$k_{cat} = V_{max} \cdot M_r / [\text{kinesin}]_{assay}$$

$[\text{kinesin}]_{assay}$: concentration of kinesin heads in the assay in g/l
$M_r$: molecular mass of a single kinesin chain in g/mol

Insertion of the corresponding numbers for the ATPase assay yields:

$$k_{cat} = 2.1436 \cdot 10^{-4} \cdot (V_{max} \cdot M_r / [\text{kinesin}])$$

[kinesin]: concentration of kinesin heads in the preparation determined by the Bradford assay (2.3.5)

## 2.3.14 Measurements under reducing and oxidizing conditions

Kinesin constructs with a cysteine residue introduced at the N-terminus of the neck coiled-coil (P342C) were characterized in terms of ATP turnover and gliding velocity under different redox conditions [Hahlen, 2004b]:

- Buffer conditions: Without DTT (dithiothreitol) or other reducing agents cysteine residues were subject to air oxidation. Here kinesin was used in assays as it was purified.
- Reducing conditions: The formation of disulfide bridges by cysteine residues was prevented due to incubation of 10-30 $\mu$M kinesin with 2-5 mM DTT on ice for 2 h prior to measuring.
- Oxidizing conditions: Formation of disulfide bridges was induced by incubation of 10-30 $\mu$M kinesin with 0.2 mM DTNB (5,5'-dithiobis(2-nitrobenzoic acid)) for 5 min on ice before starting the assays. DTNB consists of two identical parts linked by a dithiol bond. A DTNB molecule first reacts with the thiol group of one cysteine, forming a disulfide bond and releasing the reduced half of the DTNB molecule. If there is a second cysteine in close proximity, the remaining part of the DTNB molecule reacts with the thiol group of this cysteine. This results in a disulfide bridge between the 2 cysteines and release of the second reduced half of the DTNB molecule [Riddles et al., 1983].
- "Rescue" conditions: To show reversibility of the crosslink, kinesin was first oxidized with 0.2 mM DTNB for 5 min on ice and then reduced again by incubation on ice with 5 mM DTT for 4 h before measuring.

By measuring gliding velocities (2.3.11.3) of P342C-constructs under buffer, reducing, oxidizing and rescue conditions, the respective reagents were also present in the casein buffer to keep final concentrations of reagents constant during the assay.

Microtubule-activated ATP turnover rates (2.3.13.12) of P342C-constructs were also measured under different redox-conditions, however respective reagents were not present in the reaction due to the short time period of the assay (1 min).

## 2.3.15 MantADP release measurements

### 2.3.15.1 Generation of kinesin*mantADP complexes

Kinesin was incubated with a 3 fold molar excess of mantATP (2`-(3`)-0-(N-methylanthraniloyl)adenosine-5`-triphosphate, Molecular Probes, M-12417) at room temperature for 15 min in the dark. This procedure leads to a mantADP-charged kinesin because ADP bound in the nucleotide pocket is exchanged with the mantATP that is hydrolysed quickly even in the absence of microtubules. Unbound mantATP was removed by gel filtration over a MicroSpin G-25 column (Amersham Pharmacia). The column was pre-treated with 300 µl of 1 mg/ml BSA in 12A25+ for 30 min, and then equilibrated with 5-6 column volumes of 12A25+. 50 µl of the kinesin mantATP mixture were then loaded onto the column and centrifuged with 3000 rpm for 2 min. The eluate contained the kinesin*mantADP complexes (modified after [Kallipolitou et al., 2001]).

### 2.3.15.2 Concentration of mantADP complexes

To determine the concentration of the kinesin*mantADP complexes and the efficiency of mantADP binding, an absorption spectrum was obtained from 250 to 450 nm in a spectro- photometer. The maximal protein absorption is at 280 nm and the absorption maximum of the methylanthraniloyl group (mant) at 356 nm. Therefore, the protein concentration and the concentration of mant were estimated from the absorption at 280 and 356 nm and the corresponding extinction coefficients. The extinction coefficient of kinesin was calculated using the program Peptidemap from the GCG software package, the extinction coefficient of mantATP was taken from manufacturer's information.

Extinction coefficients:
$\varepsilon_{356}$ (mant) = 5800 / M∗cm
$\varepsilon_{280}$ (NcKin$_{433}$) = 26920 / M∗cm
$\varepsilon_{280}$ (ADP) = $\pm$ 2250 / M∗cm
$\varepsilon_{280}$ (NcKin$_{433}$*mantADP) = 29170 / M∗cm

### 2.3.15.3 Stoichiometry of mantADP release

Basal ADP release of kinesin is very slow but can be strongly activated by microtubule binding [Ma et al., 1995a]. Since the mant-fluorophore has a stronger fluorescence when bound to a protein than when free in solution, the release of mantADP can be followed in the fluorimeter [Ma et al., 1997a; Kallipolitou et al., 2001].
All assays were performed in an Aminco Bowman spectrofluorimeter at 22°C in 1 ml cuvettes (Greiner, 613101) and using the ATPase buffer 12A25+. The excitation wavelength was 365 nm, the emitted fluorescence was monitored at 445 nm. 100-250 nM of kinesin*mantADP complexes were mixed with at least equimolar amounts of microtubules (0.1-1 µM final concentration), either by adding them directly or by titrating them in small aliquots. To follow the release of the mantADP from the first head, the microtubule preparations contained 2 U/ml apyrase to remove traces of ATP or GTP, that would start the reaction cycle. The decrease of fluorescence was

monitored in a time course or by noting the signal after each addition. To release the second mantADP, an excess of ATP (1-2 mM) was subsequently added.

### 2.3.15.4 Pre steady-state kinetics of the mantADP release

Pre steady-state kinetics of mantADP-release was measured in a BioLogic stopped-flow apparatus (SFM-3). All assays were performed at 22 °C in 12A25+. An excitation wavelength of 365 nm was used, emission was monitored using a narrow band interference filter transmitting light with a slit of 442 ± 10 nm.

To determine mantADP release rates, 200-300 nM of kinesin·mantADP complexes (all concentrations are final concentrations) were rapidly mixed with 0-18 $\mu$M microtubules in the presence of 2 mM ATP. For each microtubule concentration 3-5 traces were obtained and averaged. All traces could be fitted to single exponentials using the software BioKine supplied by the manufacturer (BioLogic). The resulting rates showed a hyperbolic dependence on the microtubule concentration and could be fitted to the hyperbolic equation:

$y = k_{max} \cdot [Mt] / ([Mt] + K_{D,MT}) + B$

$k_{max}$: maximal ADP-release rate
$K_{D,MT}$: microtubule binding constant
B: background signal

# 3 Results

## 3.1 Sequence comparison of the neck and hinge domains of animal and fungal kinesins

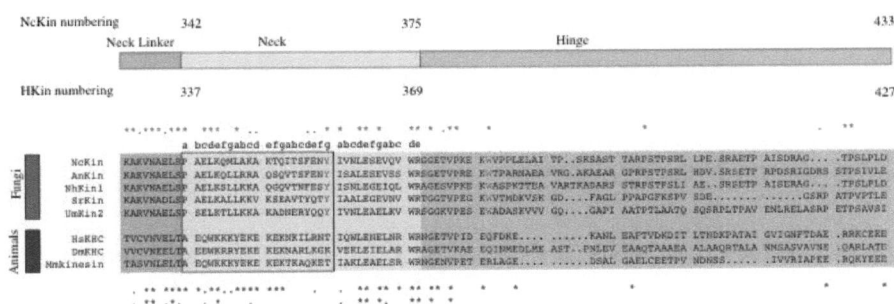

**Fig. 3.1: Sequence alignment of several fungal and animal kinesins in the neck and hinge regions.**
The distinct sequence patterns of fungal and animal neck regions are clearly visible in the first three heptads of the neck coiled-coil (heptad structure indicated above the alignment). The so called "EKEKEK" motif in animal kinesins is not present in the fungal sequences, however there are several strictly conserved motifs or residues that are specific for fungal kinesins, highlighted in red. Sequence conservation is indicated by asterisks for amino acid identity and dots for similarity, using red symbols for sequence conservation within fungi (above the alignment), blue symbols for conservation within animals and black symbols for conservation in both kingdoms (below the alignment). Kinesins from NcKin: *N. crassa*, AnKin: *A. nidulans*; NhKin1: *N. haematococca*; SrKin: *S. racemosum*; UmKin2: *U. maydis* kinesin 2; HsKHC: *Homo sapiens* KHC; DmKHC: *Drosophila melanogaster* KHC; MmKin: *Mus musculus*

The sequence of the neck domain is strongly conserved within the two kingdoms of animals and fungi, although considerably different from each other (Fig. 3.1). A striking characteristic in animal kinesins is the presence of charged residues in *a* and *d* positions in the hydrophobic core of the coiled-coil, like Tyr344, Glu347 and Asn351 in the human kinesin sequence. These residues probably destabilize a coiled-coil conformation [Tripet *et al.*, 1997], indicating that dimerization may not be the only function of the neck domain.

Fungal kinesins also display a highly conserved sequence pattern in the neck domain that differs strongly from the animal sequence, primarily in the first three heptads. The so-called "EKEKEK" motif and the coiled-coil destabilizing residues are not present in the fungal sequences, however there are several strongly conserved residues that are specific for fungi, for example Phe/Tyr358 or Tyr362.

The fungal hinge regions contain several proline and glycine residues and thus are not predicted to form a coiled-coil structure [Grummt et al., 1998b]. There is very poor sequence conservation in the hinge domain in conventional kinesins, with one conspicuous exception: filamentous fungi possess a strictly conserved residue, Trp384, not present in animal sequences.

## 3.2 Characterization of point mutations in the neck domain

In fungal necks, a highly conserved tyrosine is found in position 362 that is not present in animal kinesins. The strict conservation suggests an important role of this amino acid for the motor; therefore a mutagenesis approach was used to study the function of Y362.

### 3.2.1 Design of the point mutants

The point mutations were introduced in two different C-terminal truncated backgrounds, since these are more stable and can be bacterially expressed easier than the full-length NcKin (NcKin$_{928}$) (A. Kallipolitou & G. Woehlke, personal communication). Moreover, the shortened constructs do not contain the tail domain that inhibits motor activity [Coy et al., 1999; Friedman et al., 1999; Stock et al., 1999; Seiler et al., 2000].

The NcKin$_{433}$ construct ends at amino acid 433 behind the hinge region. It was shown to be dimeric and to display similar ATPase and motility properties as the full-length protein [Kallipolitou et al., 2001]. The NcKin$_{383}$ construct contains the entire neck domain and the first 8 amino acids of the hinge. Interestingly, these constructs are monomers, indicating that the NcKin neck domain is not sufficient for dimerization [Kallipolitou et al., 2001]. Thus, the NcKin$_{433}$ and NcKin$_{383}$ constructs allow for comparison of Tyr362 exchange mutants in single and double-headed kinesin backgrounds.

However, truncated constructs do not readily adhere to glass surfaces, which is important for the performance of motility assays (2.3.1.1). To overcome this problem, a so-called "cys-tag" was appended C-terminal to the constructs. This short stretch of 9 amino acids contains a reactive cysteine residue [Funatsu et al., 1997], that can be biotinylated using biotin-maleimide and in turn binds to streptavidin-coated coverslips (2.3.11.2). Previous studies showed that the C-terminal cyst-tag does not change the catalytic and motile properties of the motor [Kallipolitou et al., 2001].

To test the role of the conserved tyrosine 362 for motor function, it was exchanged against lysine, the corresponding residue in *Drosophila* conventional kinesin (NcKin$_{433/383}$Y362K). For control if the observed effects are really due to the lack of the tyrosine or to the introduced lysine, Y362 was also replaced with cysteine (NcKin$_{433}$Y362C) and phenylalanine (NcKin$_{433}$Y362F). Phenylalanine is very similar to tyrosine, but lacks one hydroxyl-group. Thus it allows for testing whether the observed effects depend on the aromatic ring of tyrosine. The second conserved aromatic amino acid in the fungal neck domain is F/Y359. To investigate whether the phenotype is specifically caused by Y362 or by the absence of any aromatic residue in the neck region, F359 was also exchanged against cysteine (NcKin$_{433}$F359C). Figure 3.2 summarizes all point mutants and the corresponding wild-type constructs that were characterized in this study.

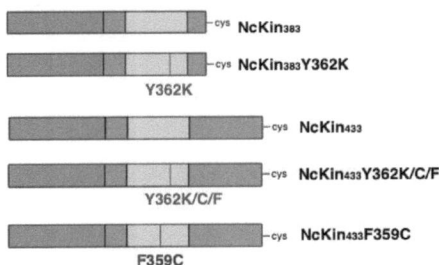

**Fig. 3.2: Point mutants and corresponding wild-type constructs.**
NcKin$_{433}$ constructs end at amino acid 433 behind the hinge region. NcKin$_{383}$ constructs contain the entire neck domain and the first 8 amino acids of the hinge domain. To facilitate biotinylation a cys-tag is appended C-terminal to the constructs. Protein domains are depicted in different colours: head: blue; neck-linker: magenta; neck: yellow; hinge: green

### 3.2.2  Isolation and biotinylation of NcKin constructs

All NcKin constructs could be expressed in *E. coli* (2.3.3) and were purified by a single ion exchange chromatography step (2.3.4.1). Protein preparations were generally very clean (Fig. 3.3, A), with concentrations ranging from 0.4 to 2.5 mg/ml (10 – 55 $\mu$M) and a total yield of about 1.5 - 7.5 mg protein.

To perform microtubule gliding assays, the cysteine-tagged versions of the NcKin433-constructs has to be biotinylated using biotin-maleimide (2.3.10). Active kinesin was isolated after the labelling reaction in a microtubule binding and release step [Vale *et al.*, 1985]. To bind the motor strongly to the microtubule, kinesin was incubated with microtubules in the presence of AMP-PNP, a non-hydrolysable ATP analogue. After centrifugation the pellet was resupended in ATP-buffer, thereby releasing active motor proteins. Because of minor microtubule-depolymerization during this procedure, a small amount of tubulin was always present in the release fraction (Fig. 3.3, B). Since the biotinylated constructs were only used for gliding assays but not for microtubule-dependent ATPase measurements, the tubulin contamination was neglected. The typical yield after biotinylation and microtubule affinity purification was 20-30% of the original kinesin preparation.

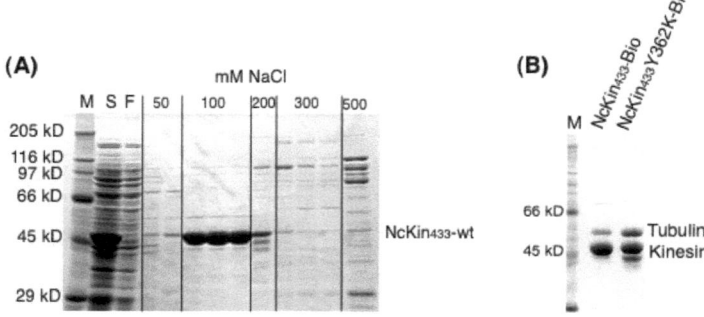

**Fig. 3.3: Isolation and biotinylation of NcKin$_{433}$-constructs.**
(A) Coomassie-stained SDS gel with representative samples of a NcKin$_{433}$-wt preparation from *E.coli*. Most of the protein eluted at 100 mM NaCl with more than 95% purity. M: protein size standard; S: supernatant after ultracentrifugation of the bacterial extract F: flow through the SP-sepharose column; 50-500 mM NaCl: eluted fractions with increasing salt concentrations
(B) Coomassie-stained SDS-gel with biotinylated NcKin$_{433}$-wt and NcKin$_{433}$Y362K after a microtubule binding and release step. In both preparations a tubulin contamination can be seen at about 55 kD, caused by microtubule-depolymerization during the release step.

### 3.2.3 Motility

The motility of dimeric NcKin constructs was tested in a microscopic gliding assay. The biotinylated constructs were bound at high density onto streptavidin coated coverslips (about 1 µg, corresponding to app. 28000 motors/µm$^2$), causing the microtubules to be transported by many motor molecules that act together on one microtubule (2.3.11.2). Under these so-called multiple motor conditions the 433-constructs showed sustained and uniform movement. The NcKin$_{433}$-wt produced fast gliding (2.2 ± 0.2 µm/s, Table 3.I), similar to the full-length NcKin (2.6 ± 0.3 µm/s, [Kallipolitou *et al.*, 2001]). The point mutant NcKin$_{433}$Y362K showed a dramatically decreased gliding velocity (1.0 ± 0.2 µm/s), about 45% of the wild-type construct. Surprisingly, the NcKin$_{433}$Y362F mutant also exhibited a dramatic decrease in velocity compared to the wild-type

(1.6 ± 0.3 μm/s), although phenylalanine is very similar to tyrosine, lacking only one OH-group. The gliding velocities of NcKin$_{433}$Y362C and NcKin$_{433}$F359C mutants could not be determined since the introduced cysteine residue made specific biotinylation impossible.

### 3.2.4 ATPase measurements

#### 3.2.4.1 Steady-state ATPase activity

To test whether this dramatic decrease of velocity was due to impaired ATP hydrolysis, the steady-state ATPase activities of the constructs were tested using a coupled enzymatic assay (2.3.13.2) [Hackney, 1994b]. Since kinesin is a microtubule-activated ATPase, the steady-state rates were measured with different microtubule concentrations and the resulting hyperbolic plots were analyzed using the Michaelis-Menten equation. The resulting maximal turnover rate per kinesin head represents the $k_{cat}$ value of the kinesin construct. The $K_{0.5,MT}$ value gives the microtubule concentration that leads to half maximal activation of the construct. It corresponds to the $K_m$ value for Michaelis-Menten enzymes, however, since conventional kinesin is highly processive, the $K_{0.5,MT}$ is much lower than expected from its microtubule binding affinity. The $K_{0.5,MT}$ therefore is related to both the processivity and the microtubule affinity of a kinesin construct.

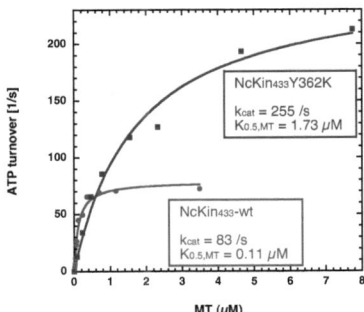

**Fig. 3.4: Microtubule-stimulated ATPase activity of NcKin$_{433}$-wt and NcKin$_{433}$Y362K.**
Steady-state rates of one example experiment are plotted versus microtubule concentration. A hyperbolic equation was fitted to the data points, yielding the maximal turnover rate $k_{cat}$ and the half-maximal saturation concentration of microtubules $K_{0.5,MT}$.
Fitting equation:      $y = k_{cat} \cdot [MT] / (K_{0.5,MT} + [MT]) + B$      B: background activity
The averages of at least two independent measurements are shown in Table 3.I.

NcKin$_{433}$-wt exhibited a k$_{cat}$ of 74 ± 8 /s (Table 3.I), similar to the full-length protein [Kallipolitou et al., 2001]. The NcKin$_{433}$Y362K point mutant showed an extremely accelerated ATP turnover of 262 ± 7 /s (Table 3.I). Similar values were obtained for the NcKin$_{433}$Y362C mutant with 176 ± 52 /s (average of 212 /s and 139 /s for two independent preparations), confirming that it is the absence of the tyrosine residue that results in a mutant phenotype. The Y362 mutants also showed significantly increased K$_{0.5,MT}$ values (for example 1.73 $\mu$M for NcKin$_{433}$Y362K versus 0.11 $\mu$M for NcKin$_{433}$-wt; Fig. 3.4). However, the K$_{0.5,MT}$ values varied strongly with different microtubule preparations, so this constant was not considered further.

The truncated NcKin$_{383}$-wt construct displayed very slow steady-state ATPase (24 ± 4 /s), as previously reported [Kallipolitou et al., 2001]. Also in this background the Y362K mutation lead to a strong acceleration of the ATP turnover (k$_{cat}$ = 157 /s). Interestingly, the NcKin$_{433}$Y362F mutant showed a slow turnover (32 ± 15 /s) comparable to that of the NcKin$_{383}$-wt construct. Thus, in contrast to the motility assays, the effect of the Y362 mutation on the ATP turnover depends on the aromatic ring of the tyrosine (rather than the OH group).

To test whether a pi-pi stacking interaction of Y362 and F359 is required for proper neck functionality, the other aromatic residue in the conserved fungal neck sequence, F359, was exchanged against cysteine. The k$_{cat}$ of the NcKin$_{433}$F359C construct varied from 44 to 88 /s, depending on the preparation, and thus did not show the extreme acceleration seen in the Y362 mutants.

### 3.2.4.2 Basal ATPase activity

To elucidate whether the acceleration of ATP hydrolysis in the Y362K point mutant was due to an intrinsically higher activity of the construct, the basal ATPase activities of wild-type and mutant NcKin$_{433}$ were measured (Table 3.I). Since the ATPase rate of kinesin in the absence of microtubules is too low to be measured in the coupled enzymatic assay, the basal rates were determined by following the phosphate release via colorimetric detection (2.3.13.1) [Geladopoulos et al., 1991]. The basal ATP turnover differed less than those of wild-type constructs of different lengths (0.02-0.03 /s, [Kallipolitou et al., 2001]). Therefore the catalytic mechanism of the Y362K mutant seems to be unchanged, but the stimulation by microtubules is much more pronounced than in the case of the wild-type kinesin.

**Table 3.I: Gliding velocities and ATPase activities of NcKin wild-type and point mutants.**

| Construct | Gliding velocity [$\mu$m/s] | Steady-state activity $k_{cat}$ [1/s] | Basal activity $k_0$ [1/s] |
|---|---|---|---|
| NcKin$_{433}$-wt | 2.2 ± 0.2 | 74 ± 8 | 0.017 ± 0.001 |
| NcKin$_{433}$Y362K | 1.0 ± 0.2 | 262 ± 7 | 0.027 ± 0.000 |
| NcKin$_{433}$Y362C | -[a] | 176 ± 52 | n.d. |
| NcKin$_{433}$Y362F | 1.6 ± 0.3 | 32 ± 15 | n.d. |
| NcKin$_{433}$F359C | -[a] | 66 ± 31 | n.d. |
| NcKin$_{383}$wt | 0.8 ± 0.2[b] | 24 ± 4[b] | n.d. |
| NcKin$_{383}$Y362K | n.d. | 157 ± 0 | n.d. |

The values are averages of at least two independent measurements.
[a] Biotinylation impossible
[b] Data from Kallipolitou *et al.* (2001).

## 3.2.5 Oligomerization state

A monomeric NcKin construct that contains only the head domain and lacks the neck and all subsequent domains (NcKin$_{343}$) behaves very similar to the NcKin-Y362 point mutants; it also moves very slowly (0.65 $\mu$m/s) but hydrolyses ATP at a high rate ($k_{cat}$ = 260 /s) [Kallipolitou *et al.*, 2001]. Therefore, the oligomerization state of the point mutants was determined by measuring the Stokes-radii ($r_{stokes}$) and Svedberg-constants ($S_{w,20}$) of the constructs via gel filtration and sucrose density centrifugation (Fig. 3.5 and 3.6) (2.3.6). The derived molecular masses were subsequently compared to the molecular masses calculated from the protein sequences to assign the oligomerization state.

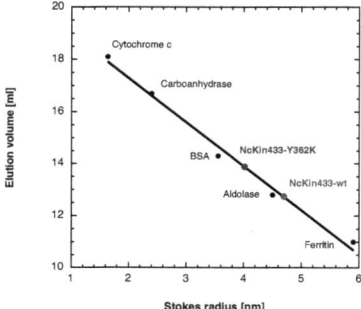

**Fig. 3.5: Gel filtration experiment with NcKin$_{433}$-wt and NcKin$_{433}$Y362K.**
The elution volumes of the reference proteins are plotted versus the Stokes radii: cytochrome c (1.64 nm), carboanhydrase (2.4 nm), BSA (3.55 nm), aldolase (4.5 nm) and ferritin (5.9 nm). Using this as a standard curve, the Stokes radii of the kinesin constructs could be determined, as depicted here for NcKin$_{433}$-wt (4.7 nm) and NcKin$_{433}$Y362K (4.0 nm).

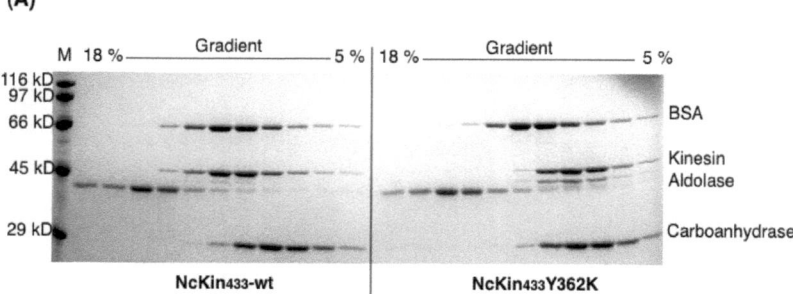

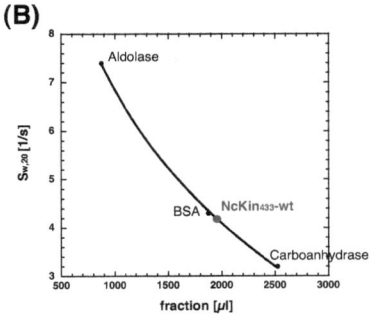 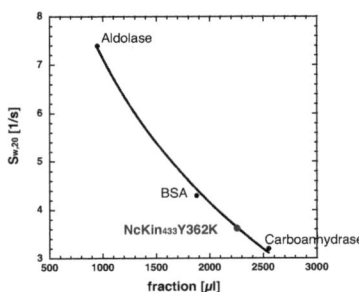

**Fig. 3.6: Sucrose density centrifugation with NcKin$_{433}$-wt and NcKin$_{433}$Y362K.**
(A): Coomassie-stained SDS-gel with gradient fractions after ultracentrifugation. The positions of the reference proteins aldolase, BSA and carboanhydrase and the kinesin sample (left: NcKin$_{433}$-wt, right: NcKin$_{433}$Y362K) in the gradient were determined by densitometric analysis using the NHI image software.
(B): The fraction volumes of the reference proteins in one example experiment were plotted against their sedimentation coefficients (aldolase: $S_{w,20} = 7.4$, BSA: $S_{w,20} = 4.3$, carboanhydrase: $S_{w,20} = 3.2$) and used as a standard to determine the sedimentation coefficient of the kinesin constructs, as depicted for NcKin$_{433}$-wt ($S_{w,20} = 4.3$) and NcKin$_{433}$Y362K ($S_{w,20} = 3.3$).

The deduced molecular masses of NcKin wild-type and mutant constructs in comparison with their calculated masses are shown in Table 3.II. In contrast to the wild-type, the point mutants NcKin$_{433}$Y362K and -Y362C clearly appeared as monomers (derived molecular masses: 54.5 kD and 56.5 kD versus calculated mass: 47.9 kD). It is noteworthy that the NcKin$_{433}$Y362F variant was also monomeric (53.5 kD versus 47.9 kD), suggesting a significant contribution of the OH-group to the stabilization of the neck coiled-coil. Unlike for the Y362 mutants, the dimerization of NcKin$_{433}$F359C was not affected (also in presence of 1 mM DTT), so apparently the dimerization depends specifically on Tyr362.

The shorter construct NcKin$_{383}$-wt failed to dimerize (49.5 kD versus 42.7 kD), consistent with earlier studies [Kallipolitou, 2002]. As expected, a monomeric state was also obtained with the NcKin$_{383}$Y362K construct (47.6 versus 42.7 kD).

**Table 3.II: Oligomerization of NcKin wild-type and point mutants.**

| Construct | $S_{w,20}$ [1/s] | $r_{stokes}$ [nm] | Derived molecular mass [kD] | Predicted mass from sequence [kD] | Oligomerization state |
|---|---|---|---|---|---|
| NcKin$_{433}$-wt | 4.3 | 4.7 | 83.4 | 47.9 | Dimer |
| NcKin$_{433}$Y362K | 3.3 | 4.0 | 54.5 | 47.9 | Monomer |
| NcKin$_{433}$Y362C | 3.6 | 3.8 | 56.5 | 47.9 | Monomer |
| NcKin$_{433}$Y362F | 3.5 | 3.7 | 53.5 | 47.9 | Monomer |
| NcKin$_{433}$F359C | 4.7 | 4.3 | 83.4 | 47.9 | Dimer |
| NcKin$_{383}$-wt | 4.0 | 3.0 | 49.5 | 47.9 | Monomer |
| NcKin$_{383}$Y362K | 3.6 | 3.2 | 47.6 | 47.9 | Monomer |

Values are averages from at least two independent measurements.

## 3.2.6 MantADP release of NcKin-Y362K

### 3.2.6.1 Stoichiometry of the microtubule-activated mantADP release

The microtubule-dependent ADP release of kinesin constructs was followed via the fluorescent ADP analogue *N*-methylanthranoyl-ADP (mantADP), which exhibits a higher fluorescence when bound to a protein than when free in solution [Ma *et al.*, 1997a; Ma *et al.*, 1997b]. NcKin was incubated with mantADP yielding a kinesin-mantADP complex since ATP is hydrolysed quickly, even in the absence of microtubules (2.3.15.1).

To validate the method, the stoichiometry of the mantADP release was assayed (2.3.15.3). NcKin$_{433}$-mantADP complexes were mixed in the fluorimeter first with microtubules and then with ATP. Figure 3.7A shows the change in the fluorescence signal when using wild-type NcKin$_{433}$-mantADP complexes. On mixing with an excess of microtubules, the signal decreased to about half of its initial value (1.0), reflecting the release of mantADP from the first head (half-site catalysis). To prevent traces of ATP or GTP from starting off the cycle, 2 U/ml apyrase were present in the microtubule fraction. When ATP was added, the fluorescence signal decreased to a background level (0.0) of free mantADP in solution due to the release of mantADP from the second head. Figure 3.7B shows a titration experiment comparing NcKin$_{433}$-wt-mantADP and NcKin$_{433}$Y362K-mantADP complexes. Microtubules were added in small aliquots until stoichiometric amounts were reached. For both constructs the fluorescence decreased markedly. In the case of NcKin$_{433}$wt-mantADP, the signal reached a plateau, reflecting the binding of the first head. The observation that

>50 % of the mantADP is released may be due to a small extent of mantADP dissociation from the second head [Ma et al., 1997a; Hackney, 2002].

In agreement with the monomeric state of the mutant protein, addition of microtubules to mantADP-labelled NcKin$_{433}$Y362K led to an ATP-independent decrease of the fluorescence signal to background levels, and was not altered further by addition of ATP.

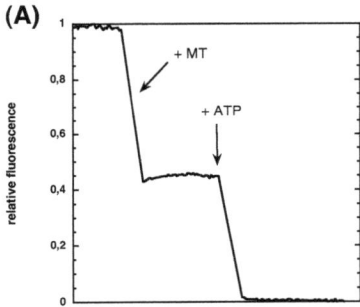 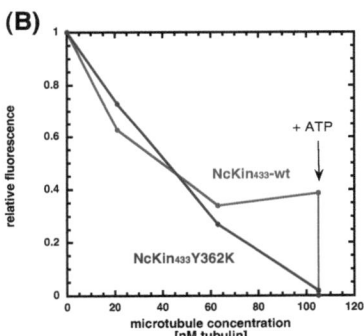

**Fig. 3.7: Stoichiometry of the microtubule activated mantADP release from NcKin$_{433}$-wt and NcKin$_{433}$Y362K.**
NcKin (100-250 nM) was mixed with microtubules to follow ADP release.
**(A):** At equimolar amounts of microtubules, the signal reaches a plateau at 0.4, indicating the release of ~60% of the mantADP. After adding 1 mM ATP, the residual mantADP is released and the signal decreases to the background level of free mantADP in solution.
**(B):** Titration of NcKin$_{433}$-wt-mantADP (red line) and NcKin$_{433}$Y362K-mantADP (blue line) complexes with microtubules. In contrast with the wild-type, the mutant releases all its mantADP upon microtubule binding without ATP, indicating uncoupled heads.

### 3.2.6.2 Pre steady-state kinetics of mantADP release

The NcKin$_{383}$Y362K mutant is of particular interest since the corresponding wild-type construct is also monomeric, allowing kinetic comparisons that are unaffected by inter-head coordination of the wild-type version. To characterize the kinetic step that is affected by Y362 in more detail, stopped-flow assays were performed on NcKin$_{383}$ wild-type and mutant variants.

The kinetics of effective microtubule binding followed by mantADP release was assessed by mixing NcKin$_{383}$-mantADP complexes with varying microtubule concentrations in the stopped-flow instrument in the presence of 2 mM ATP (Fig. 3.8) (2.3.15.4). The traces could be fitted by single-exponential equations. For the wild-type case the apparent rates $k_{obs}$ extrapolated to a maximum $k_{max}$ of 29 ± 3 /s (Fig. 3.8A, Table 3.III). This is similar to the $k_{cat}$ (24 /s), confirming the ADP release to be the rate-limiting step in the reaction cycle of a monomeric kinesin [Ma et al., 1995b]. The microtubule-stimulated mantADP release of mutant NcKin$_{383}$Y362K also obeyed single-exponential dependence at all microtubule concentrations, but extrapolated to a 6.6-fold faster $k_{max}$ (191 ± 34 /s). The $k_{cat}$ of this construct (157 /s) is accelerated by a factor of 6.5, suggesting that Y362 down-regulates the ATP turnover of the motor core at the level of the ADP release step. The dissociation constants for productive microtubule binding $K_{D,MT}$ were 3.9 ± 1.4 $\mu$M (wild-type) and 7.7 ± 2.2 $\mu$M (Y362K), respectively.

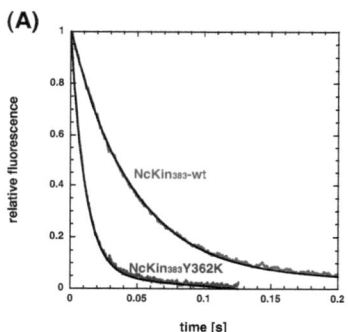 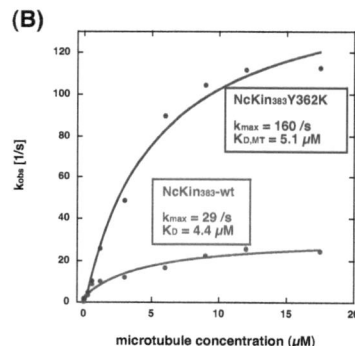

**Fig. 3.8: Pre steady-state kinetics of mantADP release from NcKin$_{383}$-mantADP complexes.**
(A): Two examples of stopped-flow traces monitored by mixing 0.2-0.3 $\mu$M NcKin$_{383}$-wt-mantADP complexes (red curve) and NcKin$_{383}$Y362K-mantADP complexes (blue curve) with 6 $\mu$M microtubules in presence of 2 mM ATP. The traces are averages of 3 to 5 single transients. The black curves are fits to a single exponential equation: y = B + A · exp (-k · t)      A: amplitude; B: background
(B): The observed rates $k_{obs}$ for one example of wild-type (red) and one example of the Y362K mutant (blue) plotted against the microtubule concentration show a hyperbolic dependence. Fitting with a hyperbolic equation gives the maximal rate $k_{max}$ and the dissociation constant for productive microtubule binding $K_{D,MT}$. The averages of three independent measurements are shown in Table 3.III.

**Table 3.III: Pre steady-state kinetics of the mantADP release**

| NcKin$_{383}$-wt | | NcKin$_{383}$Y362K | |
|---|---|---|---|
| $k_{max}$ [1/s] | $K_{D,MT}$ [$\mu$M] | $k_{max}$ [1/s] | $K_{D,MT}$ [$\mu$M] |
| 29 ± 3 | 3.9 ± 1.4 | 191 ± 34 | 7.7 ± 2.4 |

## 3.3 Characterization of NcKin-HsKin chimeric mutants

Sequence comparison of fungal and animal kinesin-1 revealed high degree of sequence conservation within the motor core but distinct and characteristic sequence patterns within the neck and the hinge regions (Fig. 3.1). The first part of this work addressed the function of the strictly conserved fungal residue Tyr 362 using a mutagenesis approach. In the second part, the role of the specific neck and hinge regions for the velocity, the mechanism of movement and/or the regulation of fungal kinesins was examined using chimeric mutants, in which the entire neck and hinge domains of NcKin were replaced by human kinesin or artificial sequences.

### 3.3.1 Design of constructs

All kinesin constructs used in this work were C-terminally truncated. Monomeric NcKin constructs consist of amino acids 1-344, comprising head and neck-linker domains (NcKin$_{344}$, Fig. 3.9). The corresponding HsKin construct includes amino acids 1-339 in the human kinesin sequence (HsKin$_{339}$). These constructs represent the "minimal motor unit" that is required for ATPase activity and mobility of the kinesin motor (1.2.1.1).

Dimeric constructs additionally contain the neck and hinge domains (amino acids 1–433 for NcKin). As mentioned earlier for the point mutants, the truncation does not affect motility and kinetic properties of the motor. However, such constructs do not readily adhere to glass surfaces. In contrast to the point mutants that were biotinylated and bound onto streptavidin coated coverslips to perform motility assays, another approach was used availing unspecific adhesion of the human kinesin tail domain (2.3.11.3). Thus, all NcKin$_{433}$ constructs were cloned in a second background, appending a part of the human stalk region (hTail: aa 432-546) C-terminal of the fungal hinge domain (NcKin$_{hTail}$) [Hahlen, 2004b]. The hTail-versions have the added advantage that the coiled-coil hTail [de Cuevas *et al*., 1992] readily promotes dimer formation even in mutants that prevent neck dimerization. The hTail–appendix did not contain the IAK-motif responsible for the tail inhibition [Seiler *et al*., 2000] in order to prevent intramolecular motor regulation. The corresponding human kinesin construct used in this work was HsKin$_{546}$.

Four different chimeric constructs were designed to study the importance of the specific hinge and neck domain for kinesin function (Fig. 3.9). In one, the entire hinge domain from position 380 to 433 was replaced by the human kinesin sequence (NcKin_379). All kinesin hinge regions show very poor sequence conservation, however there is one strictly conserved residue, W384, in the fungal hinge that is not present in the human (or any other animal kinesin-1) sequence. To assess the function of this fungal specific residue, a second hinge-chimera was generated with the human sequence starting 5 residues further C-terminal, retaining W384 (NcKin_384).

The remaining chimeric mutants address the importance of the fungal neck domain. In one construct the entire neck and hinge regions were replaced by human sequences (NcKin_340), thus containing only the NcKin motor core and neck-linker domains. In a second construct the fungal hinge domain is retained replacing only the neck (NcKin_340-79).

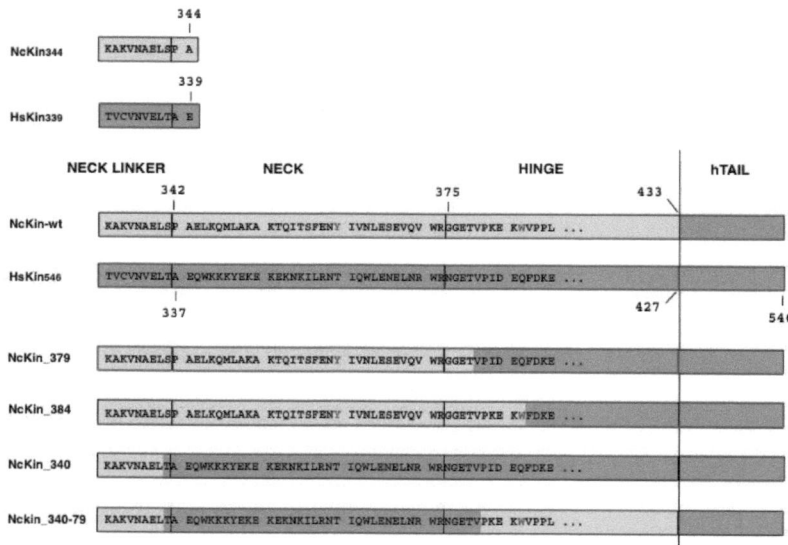

**Fig. 3.9: Design of truncated and chimeric constructs.**
Constructs are illustrated schematically with neck-linker, neck and hinge domains depicted as black-bordered boxes. NcKin sequences are shaded in light-pink, human kinesin sequences in light green. The specific neck and hinge sequences are denoted in letters, with the conserved fungal residues Y362 and W384 highlighted in red. All dimeric constructs despite HsKin$_{546}$ were generated as two versions: one ending at amino acid 433 (NcKin$_{433}$), the other version with the hTail appended C-terminal to the hinge domain (NcKin$_{hTail}$).

### 3.3.2 Temperature dependence of NcKin and HsKin

To examine whether the different catalytic and motile activities of fungal and animal kinesins are due to different temperature optima, the steady-state ATP turnover and microtubule gliding velocity was determined at different temperatures for NcKin and HsKin constructs (Table 3.IV).

For the ATPase measurements the monomeric constructs $NcKin_{344}$ and $HsKin_{339}$ were used since these constructs comprise the catalytic motor core and the neck-linker, but lack all further C-terminal regions that could possibly influence ATP turnover or temperature dependence of the head. Steady-state turnover rates were measured in a temperature range from 15°C to 35°C (Fig. 3.10, A). At room temperature (22 °C), the catalytic rates for HsKin (61 /s) and NcKin (264 /s) agreed well with previous measurements [Ma *et al.*, 1997b; Kallipolitou *et al.*, 2001]. The resulting Arrhenius plots and activation energies are very similar for NcKin and HsKin ($\Delta E_A$ = 37 kJ/mol and 48 kJ/mol, respectively) with the NcKin $k_{cat}$ values higher than the HsKin rates at each given temperature. However, NcKin turned out to be less stable at higher temperatures since it was already getting inactivated at > 28°C, whereas the HsKin activity increased steadily up to 35°C.

Microtubule gliding activities were measured with the dimeric $HsKin_{546}$ and $NcKin_{hTail}$ versions that adhere readily to coverslips. Human kinesin velocities were obtained in a temperature range from 15°C to 40°C, whereas for the NcKin construct no microtubule attachment could be observed at temperatures > 30°C (Fig. 3.10, B). The resulting Arrhenius plots were again very similar ($\Delta E_A$ = 49 kJ/mol for $NcKin_{hTail}$ and 58 kJ/mol for $HsKin_{546}$), with the NcKin velocities being always higher than those of the human kinesin.

These results indicate an intrinsically higher motile and catalytic activity of the fungal motor compared to animal kinesin that is not due to different temperature optima.

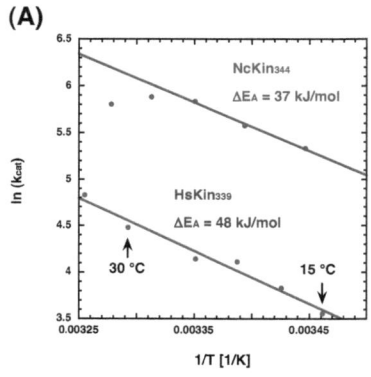
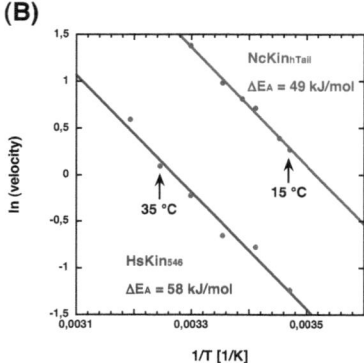

**Fig. 3.10: Temperature dependence of NcKin and HsKin.**
(A): Arrhenius plots for the dependence of microtubule-stimulated ATPase activity of NcKin$_{344}$ (red) and HsKin$_{339}$ (green) on temperature. Fitting to the Arrhenius equation revealed the Arrhenius activation energies for both constructs. Measurements were done in a temperature range from 15 to 34 °C. Above 28 °C the NcKin construct is getting inactivated; corresponding rates are therefore not considered for the fitting process.
(B): Arrhenius plots for the dependence of microtubule gliding velocity of NcKin$_{hTail}$ (red) and HsKin$_{546}$ (green) on temperature. Measurements were done in a temperature range from 15 to 40 °C. Fitting to the Arrhenius equation revealed the Arrhenius activation energies for both constructs. No microtubule attachment could be detected with NcKin$_{hTail}$ at temperatures above 30°C.

Arrhenius equation: $\ln v = A - \Delta E_A / RT$
v: reaction rate
A: Arrhenius constant
$\Delta E_A$: Arrhenius activation energy
R: gas constant
T: temperature

**Table 3.IV: Temperature dependence of microtubule-stimulated ATPase and microtubule gliding activity**

| ATPase activity | | | | Gliding velocity | | | |
|---|---|---|---|---|---|---|---|
| HsKin$_{339}$ | | NcKin$_{344}$ | | HsKin$_{546}$ | | NcKin$_{hTail}$ | |
| T [°C] | $k_{cat}$ [1/s] | T [°C] | $k_{cat}$ [1/s] | T [°C] | v [µm/s] | T [°C] | v [µm/s] |
| 15.9 | 35 | 17 | 207 | 15 | 0.29 | 15 | 1.31 |
| 18.9 | 46 | 21.5 | 264 | 20 | 0.46 | 16.5 | 1.48 |
| 22.2 | 61 | 25.3 | 341 | 25 | 0.52 | 20 | 2.04 |
| 25.4 | 63 | 28.7 | 358 | 30 | 0.80 | 22 | 2.25 |
| 30.7 | 88 | 31.9 | 331 | 35 | 1.10 | 25 | 2.67 |
| 34.2 | 125 | | | 40 | 1.81 | | 3.98 |

### 3.3.3 Gliding velocity and ATP turnover of chimeric constructs

The fast gliding velocity of NcKin requires an efficient transformation of the chemical energy derived from the high catalytic activity in the motor head into movement. In the following sections the importance of the conserved fungal neck and hinge regions for the mechanism of fast fungal kinesins is examined.

#### 3.3.3.1 Motility

Gliding velocities of chimeric constructs were determined using the hTail-versions (2.3.11.3). Microscopic gliding assays were done under multiple motor conditions (about 1 $\mu$g kinesin corresponding to 28000 motor proteins per 1 $\mu m^2$) with all constructs showing sustained and uniform movement (Table 3.V).

As expected, the wild-type NcKin motor (NcKin$_{hTail}$) displayed high gliding velocity (2.7 ± 0.3 $\mu$m/s), very similar to the biotinylated NcKin$_{433}$ version (2.2 ± 0.2 $\mu$m/s, Table 3.I) and the full-length protein (2.6 ± 0.3 $\mu$m/s, [Kallipolitou et al., 2001]). The human kinesin construct (HsKin$_{546}$) moved with a typical velocity for animal conventional kinesins (0.4 ± 0.1 $\mu$m/s), about 6-fold slower than NcKin.

The chimeric neck-constructs NcKin_340 and NcKin_340-79 were not affected in terms of motility, showing high gliding velocity similar to the wild-type (2.3 ± 0.3 $\mu$m/s and 2.2 ± 0.2 $\mu$m/s, respectively).

The only chimera that exhibited a significantly different motility from NcKin wild-type was NcKin_379 with a 70% decrease in gliding velocity (0.8 ± 0.2 $\mu$m/s). It is noteworthy, that this construct, containing the "natural" fungal neck domain, moved slower than NcKin_340 with human kinesin neck and hinge sequences. Surprisingly, the slightly longer hinge chimera that includes the conserved W384, on the other hand, exhibited fast gliding velocity (2.2 ± 0.3 $\mu$m/s), consistent with an important role of W384 for motor functionality.

#### 3.3.3.2 Microtubule-stimulated ATPase activity

Steady-state ATPase activity was determined for all constructs as 433- and hTail-versions (Table 3.V). For NcKin wild-type, the catalytic activities were indistinguishable in both backgrounds (74 ± 8 /s for NcKin$_{433}$ and 71 ± 14 for NcKin$_{hTail}$), confirming that the hTail does not influence the properties of the wild-type motor. As expected, dimeric HsKin$_{546}$ exhibited a slower ATP turnover than the corresponding NcKin construct (47 ± 2 /s).

The chimera NcKin_340 also displayed wild-type ATP turnover in both versions (74 ± 2 /s for NcKin_340$_{433}$ and 70 ± 0.6 /s for NcKin_340$_{hTail}$) and thus appeared to be completely unaffected in its motile and catalytic properties. For all other chimeras lower ATPase activities were obtained that were further decreased in the absence of the human tail sequence. NcKin_340-79 exhibited a relatively slight reduction in catalytic activity as hTail-construct (44 ± 3 /s) and a further decrease of 40 % as 433-version (27 ± 3).

The ATP turnover of the hinge-chimeras NcKin_379 and NcKin_384 was dramatically decreased in the 433-background compared to the wild-type (13 ± 3 /s and 16 ± 3 /s). In the case of Nckin_379, the catalytic activity was also strongly reduced in the hTail-version (27 ± 4 /s), whereas the slightly longer chimera NcKin_384 was much less affected in this background (56 ± 2/s). This is consistent with the gliding measurements that showed reduced motility of NcKin_379$_{hTail}$ but fast movement of NcKin_384$_{hTail}$.

**Table 3.V: Motile and kinetic characteristics of chimeric constructs**

| Construct | Gliding velocity [µm/s] | Steady-state ATPase [1/s] |
|---|---|---|
| NcKin$_{hTail}$ | 2.7 ± 0.3 | 71 ± 1 |
| NcKin$_{433}$ | n.a. | 71 ± 4 |
| HsKin$_{546}$ | 0.4 ± 0.1 | 47 ± 2 |
| NcKin_340$_{hTail}$ | 2.3 ± 0.3 | 74 ± 22 |
| NcKin_340$_{433}$ | n.a. | 70 ± 1 |
| NcKin_340-79$_{hTail}$ | 2.2 ± 0.2 | 44 ± 4 |
| NcKin_340-79hk$_{433}$ | n.a. | 27 ± 3 |
| NcKin_379k$_{hTail}$ | 0.8 ± 0.2 | 27 ± 4 |
| NcKin_379$_{433}$ | n.a. | 13 ± 3 |
| NcKin_384$_{hTail}$ | 2.2 ± 0.3 | 56 ± 2 |
| NcKin_384$_{433}$ | n.a. | 16 ± 3 |

Values are averages from at least two independent measurements.
n.a.: not applicable, constructs without hTail do not adhere on coverslips

### 3.3.4  Oligomerization state

To investigate the requirements for NcKin dimerization, the oligomerization states of the chimeric constructs were examined by gel filtration and sucrose density centrifugation. Since constructs in the hTail-background are dimeric, being linked at the C-terminus by the hTail coiled-coil, only the 433-versions were tested.

As already mentioned, the wild-type 433-construct was clearly dimeric (Tables 3.II, 3.VI). Dimeric states were also obtained for the neck chimeras NcKin_$340_{433}$ and NcKin_$340$-$79_{433}$, confirming the human neck domain to be in a two-stranded coiled-coil state [Morii *et al.*, 1997; Tripet *et al.*, 1997]. In contrast, NcKin_$379_{433}$ and NcKin_$384_{433}$ turned out to be monomeric (Table 3.VI). These observations clearly show that the NcKin neck does not readily dimerize by itself unless the specific NcKin hinge is included, in agreement with previous observations [Kallipolitou *et al.*, 2001]. They also disclose the interesting fact that the conserved tryptophan at position 384 of the hinge is not sufficient for neck dimerization.

**Table 3.VI: Oligomerization of chimeric constructs**

| Construct | $S_{w,20}$ [1/s] | $r_{stokes}$ [nm] | Derived molecular mass [kD] | Predicted mass from sequence [kD] | Oligomerization state |
|---|---|---|---|---|---|
| NcKin$_{433}$ | 4.3 | 4.7 | 83.4 | 47.9 | Dimer |
| NcKin_$340_{433}$ | 4.2 | 4.5 | 78.4 | 48.8 | Dimer |
| NcKin_$340$-$79_{433}$ | 4.4 | 4.6 | 85.5 | 48.8 | Dimer |
| NcKin_$379_{433}$ | 3.3 | 3.6 | 50.7 | 48.2 | Monomer |
| NcKin_$384_{433}$ | 3.3 | 3.6 | 50.5 | 48.3 | Monomer |

Values are averages from at least two independent measurements.

### 3.3.5  Crosslinking studies on hinge chimeras

At first sight, the monomeric state of Nkin_$384_{433}$ disagrees with the motility data, since the hTail-version of this chimera moved with high velocity (Table 3.V), suggesting the neck domain to be in a properly folded, dimeric state. Moreover, the strong decrease of ATPase activity in the 433-background compared to the hTail-background indicates that the hTail domain may affect the structural state of the neck in this mutant. However, since the hTail-versions of the chimeric mutants were used in the gliding assays, the functional state of the neck in the hTail-background had to be established. For that, a crosslinking approach was chosen.

### 3.3.5.1 Formation of crosslinks

A cysteine residue was introduced at the first position of the neck coiled-coil, replacing the endogenous P342 (Fig. 3.1), leading to a juxtaposition of the two cysteines in a dimeric kinesin neck construct [Zhou et al., 1993; Morii et al., 1997; Kallipolitou et al., 2001]. Thus, in a correctly folded coiled-coil conformation of the neck domain, the cysteines readily oxidize, forming an interstrand crosslink [Hahlen, 2004a].

The P342C mutation was introduced in the hTail-versions of the chimeric hinge-constructs and crosslink formation was tested by non-reducing SDS-PAGE (Fig. 3.11). As a negative control, wild-type NcKin$_{hTail}$ was used and showed a single band at 60 kD under all conditions, the expected electrophoretic mobility of an un-crosslinked polypeptide chain. Confirming previous results [Hahlen, 2004a], the P342C mutant displayed a clear band-shift under oxidizing conditions (DTNB or air), indicating that the crosslink is quite favourable (Fig. 3.11). Unexpectedly, the shifted band did not appear at 120 kD, the predicted mass of a dimer, but around 200 kD. Apparently, the disulfide bridge connects the two subunits in a way that severely reduces their electrophoretic mobility in the hTail-background. Crosslinking of NcKin$_{433}$P342C (without hTail) resulted in a band shift to the expected size of a dimeric kinesin (90 kD) [Hahlen, 2004a], and gel filtration and sucrose density centrifugation experiments with NcKin$_{hTail}$P342C excludes the formation of oligomers of more than two subunits (Table 3.VII).

The NcKin_384$_{hTail}$P342C chimera runs on the gel as a dimer both under oxidizing conditions and in buffer (i.e. air oxidation) (Fig. 3.11), suggesting a correctly folded neck coiled-coil. The NcKin_379$_{hTail}$P342C chimera also appeared as a dimer in the crosslinked state, despite the fact that it is monomeric in the 433-background (Table 3.VI) and exhibited slow gliding velocity as an un-crosslinked hTail construct (Table 3.V), which indicates an impaired neck dimerization. To determine whether non-reducing SDS-PAGE correctly reveals unfolded neck domains, we introduced the P342C mutation into the NcKin$_{433}$Y362K mutant that had been shown to be monomeric and slow in gliding assays (Table 3.I, II). The resulting NcKin$_{hTail}$P342C,Y362K double mutant clearly dimerizes due to the hTail-appendix, as shown in gel filtration and sucrose density centrifugation (Table 3.VII). However, no intermolecular crosslink appeared under any condition (Fig. 3.11), consistent with the neck domain being in an unfolded state. These results suggest that in contrast to the Y362K mutant the neck domains of the NcKin-379$_{hTail}$P342C dimer are in close proximity to each other to allow crosslinking at P342C, even though they are unable to provide the stability of a fully functional neck (see 4.1.2.2).

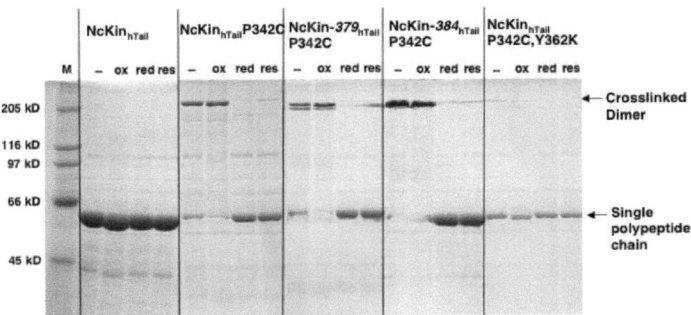

**Fig. 3.11: Crosslinking of P342C mutants.**
The mutants are fractionated on a non-reducing SDS-gel under different redox-conditions (2.3.14).--: no reagent, oxidation by air; ox: + 0.2 mM DTNB; red: + 2-5 mM DTT; res: + 0.2 mM DTNB 5', 5 mM DTT 4h.
Under oxidative conditions (air, DTNB) a band-shift to about 200 kD is visible for NcKin$_{hTail}$P342C and the chimeras NcKin_379$_{hTail}$P342C and NcKin_384$_{hTail}$P342C to a nearly complete extent, indicating an inter-chain crosslink (see also text). The crosslink is fully reversible as it is seen by the appearance of a 60 kD band (corresponding to one kinesin polypeptide chain) instead of the 200 kD band under reducing conditions (DTT). No crosslink occurred with the negative control NcKin$_{hTail}$ without P342C and with the double mutant NcKin$_{hTail}$Y362K,P342C.

### 3.3.5.2 Motility of crosslinked constructs

Crosslinking the neck domain does not affect the motility or the ATPase activity of NcKin wild-type, as shown previously [Hahlen, 2004a].

As expected, NcKin_384$_{hTail}$P342C is also a fast motor under all conditions (Table 3.VII) and thus resembles the NcKin_384$_{hTail}$ parent construct (Table 3.V). Unlike the NcKin_379$_{hTail}$ parent construct, the velocity of NcKin_379$_{hTail}$P342C was high under oxidizing conditions (2.1 ± 0.1 $\mu$m/s). This is consistent with the crosslinking behaviour described in the preceding paragraph, indicating that artificial coupling of the two heads in a dimer leads to a fast motor. Unexpectedly, it remained high at 2.0 ± 0.3 $\mu$m/s in the presence of 2 mM DTT (and decreased only slightly to 1.7 ± 0.1 $\mu$m/s with 10 mM DTT). Under rescue conditions for 4 h, the velocity was also high (1.9 ± 0.3 $\mu$m/s) and more similar to the wild-type than to the mutant with uncoupled heads. Thus, under reducing conditions NcKin_379$_{hTail}$P342C did not show the slow gliding velocity of the parent construct (~0.8 μm/sec), which one would expect in the absence of the P342C crosslink. This is in contrast to the ATPase activity that was high in the crosslinked state (64 ± 4 /s), resembling the wild-type, but low under reducing conditions (42 ± 6 /s and 30 ± 3 /s), very similar to the NcKin_379$_{hTail}$ parent construct (Table 3.VII).

**Table 3.VII: Characterization of P342C mutants**

| Construct | Gliding velocity [μm/s] | | | Steady-state ATPase [s$^{-1}$] | | | Molecular masses [kD] | | State |
|---|---|---|---|---|---|---|---|---|---|
| | ox | red | res | ox | red | res | derived | predicted | |
| NcKin$_{hTail}$P342C | 2.1 ± 0.1[a] | 2.5 ± 0.2[a] | 2.5 ± 0.1[a] | 66 ± 1[a] | 64 ± 5[a] | 63 ± 5[a] | 128.8[b] | 60.9[b] | Dimer[b] |
| NcKin$_{hTail}$P342C,Y362K | n.d. | n.d. | n.d. | n.d. | n.d. | n.d. | 106.9[c] | 60.8[c] | Dimer[c] |
| NcKin-379$_{hTail}$P342C | 2.12 ± 0.32 | 2.04 ± 0.28 | 1.90 ± 0.26 | 64 ± 4 | 42 ± 6 | 30 ± 3 | n.d. | n.d. | |
| NcKin-384$_{hTail}$P342C | 2.05 ± 0.17 | 2.12 ± 0.17 | 2.22 ± 0.23 | n.d. | n.d. | n.d. | n.d. | n.d. | |

ox: oxidation (0.2 mM DTNB, 5 min); red: reduction (2-5 mM DTT, 2 h); res: rescue (0.2 mM DTNB, 5 min; 5 mM DTT, 4h)
[a] Data from [Hahlen, 2004a]
[b]: oligomerization of NcKin$_{hTail}$P342C was tested in the oxidized state to verify formation of dimers
[c]: oligomerization of NcKin$_{hTail}$P342C,Y362K was tested to verify dimerization via the hTail domain
n.d.: not determined

Two explanations for this behaviour are conceivable and were studied experimentally. First, the P342C substitution stabilizes the α-helical conformation of the neck to allow for coiled-coil formation even in the absence of a disulfide bond; second, the P342C disulfide bridge is rather stable and cleaved only incompletely.

To address the first concern, the oligomerization states of the NcKin_379$_{433}$P342C and NcKin_384$_{433}$P342C chimeras were determined. Both displayed molecular masses corresponding to a single polypeptide chain (55.3 kD and 61.2 kD, respectively) and were therefore monomeric (Table 3.VIII), indistinguishable from the P342 wild-type situation (Table 3.VI). This argues against the first possibility. It is worth noting, however, that in the absence of DTT, two peaks appeared in the gel filtration analysis of both constructs, indicating a mixed population of dimers and monomers. Upon addition of 1 mM DTT, the dimer peak disappeared completely and the monomer peak increased proportionally (not shown). Therefore, even though the P342C mutation does not induce NcKin dimerization by itself, formation of a crosslink is apparently enhanced and already occurs in the absence of added oxidizing reagents.

**Table 3.VIII: Oligomerization of P342C mutants**

| Construct | $S_{w,20}$ [1/s] | $r_{stokes}$ [nm] | Derived molecular mass [kD] | Predicted mass from sequence [kD] | Oligomerization state |
|---|---|---|---|---|---|
| NcKin_379$_{433}$P342C | 3.6 | 3.8 | 55.3 | 48.2 | Monomer |
| NcKin_384$_{433}$P342C | 3.9 | 3.7 | 61.2 | 48.3 | Monomer |

To address the second point, gliding competition assays were performed. Since it is not feasible to maintain both crosslinked and un-crosslinked motors of NcKin_379$_{hTail}$P342C in the same preparation, because one requires oxidizing conditions and the other reducing conditions, the experiments were performed with a functionally equivalent pair of constructs: NcKin$_{hTail}$ (which possesses a coiled-coil neck and is fast), and NcKin_379$_{hTail}$ (which has a labile neck and is slow). The two were mixed and applied to a coverslip in different proportions to determine which quantity of neck-stable motor is sufficient to generate fast gliding (Fig. 3.12). The results show that as little as 10% NcKin$_{hTail}$ is able to sustain a gliding velocity of the mixture of ~2 $\mu$m/s. Therefore, the fast gliding of the oxidized NcKin_379$_{hTail}$ P342C is irreversible, most likely due to contamination with still crosslinked motors even under reducing conditions, that compromises gliding assays more than ATPase assays (Table 3.VII).

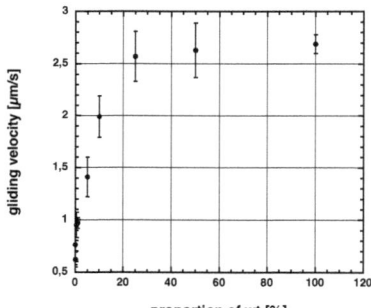

**Fig. 3.12: Competitive gliding experiments.**
Fast, wild-type NcKin$_{hTail}$ was mixed with the slow NcKin_379$_{hTail}$ chimera in variable ratios from 0-100 % and the resulting microtubule gliding velocities were determined in a multiple motor gliding experiment. 25 % of NcKin$_{htail}$ is sufficient to produce wild-type velocity. It can be estimated from the plot that a gliding velocity of 1.9-2.0 µm/s, measured in a gliding assay with NcKin_379$_{hTail}$P342C under reducing conditions, is generated by 5-10 % of the "fast" motor. Thus, approximately 5-10 % of the NcKin_379$_{hTail}$P342C preparation was still crosslinked in presence of 5 mM DTT in the gliding experiment.

### 3.3.6 Characterization of a mutant NcKin with an artificial neck domain

The preceding observations with the chimeric constructs indicate that the specific neck and hinge sequences of NcKin are not required to produce fast velocity as long as the integrity of the neck coiled-coil provides effective coupling of the two heads in a dimer (compare NcKin_340). To test this hypothesis the entire neck domain of NcKin was replaced by an artificial coiled-coil sequence (NcKin_stableNeck).

#### 3.3.6.1 Design of NcKin_stableNeck

The specific neck domain of NcKin from 342 to 374 was replaced by 5 heptads of the so-called EIEALKA sequence (Fig. 3.13). This coiled-coil model sequence was designed by Su et al., 1994 to maximize hydrophobic and ionic intra- and interchain interactions and was found to form a stable coiled-coil conformation [Su *et al.*, 1994; Tripet *et al.*, 1997].

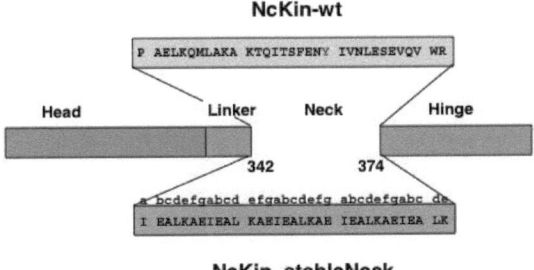

**Fig. 3.13: Design of NcKin_stableNeck.**
The entire neck region of NcKin from aa 342 to 374 was replaced by 5 heptads of the artificial coiled-coil sequence "EIEALKA" [Su *et al.*, 1994]. The heptad repeat in the wild-type sequence remained unchanged in the stableNeck-construct (indicated above the inserted sequence).

### 3.3.6.2 Biochemical characterization of NcKin_stableNeck

Similar to the chimeric constructs the NcKin_stableNeck mutant was cloned in two versions: one with (NcKin_stableNeck$_{hTail}$) and one without hTail domain (NcKin_stableNeck$_{433}$). Table 3.IX shows the biochemical properties of these mutants. Unexpectedly, the gliding velocity of the NcKin_stableNeck$_{hTail}$ construct was cut to half that of wild-type motility ($1.3 \pm 0.2 \mu$m/s), indicating that the artificial coiled-coil cannot fully substitute the specific fungal neck domain. Also the ATP turnover was slightly reduced in the hTail construct ($47 \pm 4$/s), however, in the 433-version it was similar to the wild-type ($87 \pm 5$/s). Confirming dimerization by the artificial coiled-coil domain, the NcKin_stableNeck$_{433}$ mutant clearly appeared as a dimer in gel filtration and sucrose density centrifugation experiments (72.5 kD versus 47.9 kD).

**Table 3.IX: Characterization of NcKin_stableNeck.**

| Construct | Gliding velocity [µm/s] | Steady-state ATPase [1/s] | Molecular masses [kD] | | Oligomerization state |
|---|---|---|---|---|---|
| | | | Derived | Predicted | |
| NcKin$_{hTail}$ | 2.7 ± 0.3 | 71 ± 1 | n.d. | n.d. | |
| NcKin$_{433}$ | n.a. | 71 ± 4 | 83.4 | 47.9 | Dimer |
| NcKin_stableNeck$_{hTail}$ | 1.3 ± 0.2 | 47 ± 4 | n.d. | n.d. | |
| NcKin_stableNeck$_{433}$ | n.a. | 87 ± 5 | 72.5 | 47.9 | Dimer |

n.a.: not applicable, constructs without human tail do not adhere to coverslips.
n.d.: not determined, dimerization via hTail-domain.

## 3.4 Single molecule studies

To compare the motor-mechanical properties of NcKin wild-type and NcKin_stableNeck in more detail, single molecule measurements in an optical laser trap were performed in co-operation with Johann Jaud in the group of Matthias Rief at the Technical University in Munich.

### 3.4.1 Affinity purification of motor constructs

For measurements at the single molecule level a high proportion of active motor proteins is more critical than purity of the preparation. To select for active molecules, a purification process based on a microtubule affinity step was used for motor preparation (2.3.4.3), similar to the biotinylation protocol for cys-tagged proteins (2.3.10).

After cell lysis and ultracentrifugation, the supernatant was mixed with an excess of taxol-stabilized microtubules in the presence of AMP-PNP to bind the motor proteins tightly onto the microtubule. After a wash-step, active kinesin molecules are released upon resuspending the pellet in an ATP-containing buffer in the presence of 200 mM KCl (Fig. 3.14). Preparations via microtubule affinity were typically less pure than those via a SP-sepharose column (for comparison: Fig. 3.2). However, NcKin$_{hTail}$ constructs bind exclusively to carboxylated latex beads that were used in the single molecule measurements (see below), thus contaminations were not a problem in these assays.

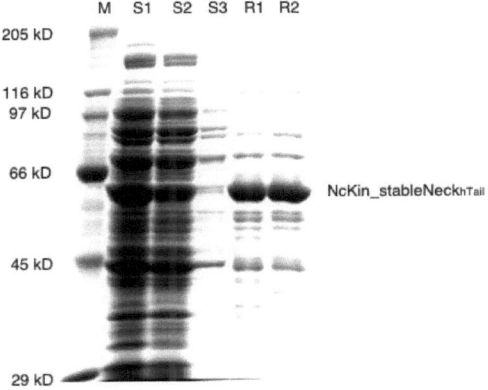

**Fig. 3.14: Representative samples of a microtubule-affinity purification of NcKin_stableNeck$_{hTail}$.**
Active kinesin molecules were purified via binding on microtubules in presence of AMP-PNP and released in presence of ATP. M: protein size standard; S1: supernatant after ultracentrifugation of bacterial cell extract; S2: supernatant after microtubule binding; S3: supernatant after washing; R1, R2: release fractions after resuspending the pellet in ATP containing buffer in presence of 200 mM KCl.

## 3.4.2   Optical laser trap

### 3.4.2.1   Principle

The technique of trapping small particles in a gradient of a focused laser beam, also referred to as "optical tweezers", was discovered in 1987 by A. Ashkin as a new tool to manipulate cells, viruses or organelles [Ashkin et al., 1987; Ashkin et al., 1990]. The possibility to detect very small forces of several pN and a spatial resolution in the nm range makes this technique also a powerful method to study single, force-producing molecules like kinesin motor proteins [Block et al., 1990].

Optical tweezers are based on the fact that electromagnetic radiation adsorbed or refracted by an object exerts a force onto this object. Under certain optical conditions like a strongly focused laser beam, dielectric particles can be trapped in a position that lies near the centre of the laser beam. Usually, small beads from silica or latex material with a typical diameter of 0.2 – 1 μm are used in laser trap experiments. Onto these, motor molecules are adsorbed at low density and delivered onto a fixed microtubule by using the laser beam as optical tweezers. By detecting the position of the bead, the movement and force production of a single motor molecule can be measured [Svoboda et al., 1993].

### 3.4.2.2 Experimental setup

All single molecule measurements were carried out in an optical laser trap in combination with a fluorescence microscope built by Anabel Clemen and Johann Jaud in the group of Matthias Rief. A scheme of the experimental setup is shown in figure 3.15.

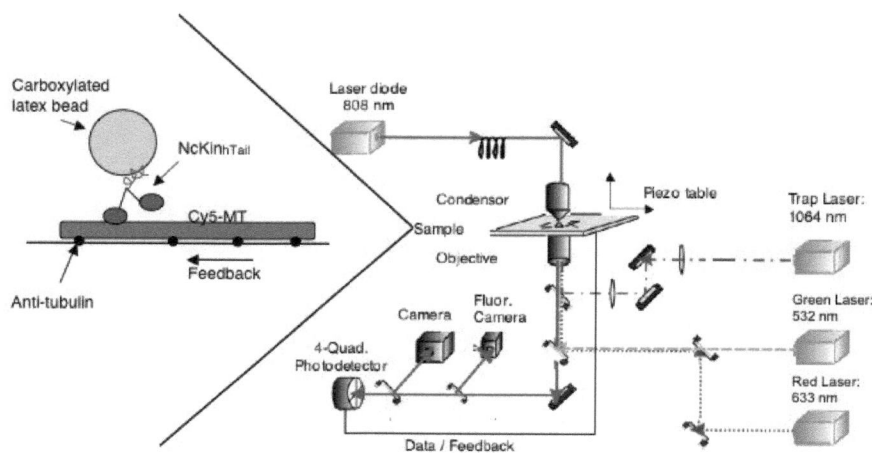

**Fig. 3.15: Schematic representation of the optical laser trap.**
The inset shows an enlarged illustration of the sample containing the kinesin molecules bound on latex beads and microtubules immobilized on the coverslip. See text for further explanations.

To minimize deleterious effects of laser radiation on the motors, infrared laser light with 1064 nm wavelength was used for the trapping laser, since light adsorption by biological molecules is smallest in this range. The beam of the trapping laser was focused via an objective onto the sample represented by a flow chamber that contained the beads and the immobilized microtubules. The position of the bead was detected via a quadrant photo diode with a red laser diode (808 nm) used for bead illumination. This laser operated with low power under the laser threshold, providing just a strong monochromatic light source.

Fluorescently labelled microtubules (Cy5-MT) were visualized in the combined fluorescence microscope using the so-called TIRF method (for *total* *internal* *reflection* *fluorimetry*) [Funatsu *et al.*, 1995]. For that, red laser light (633 nm) was focused via the objective onto the flow chamber in an angle greater than the critical angle for total reflection. Thus, almost all of the light was reflected, however, within the so-called evanescent field, a very small layer at the glass-water interface of the chamber was illuminated (about 150 nm). Thereby the background signal was strongly reduced compared to conventional fluorescence microscopy. The apparatus provided in addition a green laser for the excitation of other fluorescent dyes, but was not used in these experiments.

### 3.4.2.2.1 Calibration of the optical trap

When a bead is positioned off-centre of the laser beam, a force is exerted onto the bead that for small displacements is directly proportional to the distance between the bead and the trap. Thus, the laser trap acts like an elastic spring and by determining the spring constant it is possible to measure the force a single motor molecule can produce by moving the bead out of the trap centre. There are several methods to calibrate an optical trap. In this work the spring stiffness was determined by measurement of the thermal fluctuation of an unbound, trapped bead in one dimension [Svoboda et al., 1993]. This is a fast and easy procedure, allowing calibration of every bead that is trapped during one experiment. The typical stiffness for single molecule measurements was 40 – 70 fN/nm.

### 3.4.2.2.2 Long-range feedback system

As previously shown in single molecule TIRF assays, NcKin is a highly processive motor that is able to move more than 3 $\mu$m along the microtubule lattice before dissociating [Lakamper et al., 2003]. As mentioned above, while moving the bead out of the trap centre, a force is applied on the motor that increases linearly with the distance the motor is moving. At some point the maximum force of a single motor molecule will be reached, causing the motor to stall and subsequently to dissociate from the microtubule (maximum force = stall force). Thus, using this setup the stall force of single kinesin molecules can be determined [Svoboda et al., 1993].

To measure the maximum run length of a kinesin motor under constant forces an additional feature, called the long-range feedback system, was integrated in the optical trap. When the motor reaches a previously selected force between 0.7 to 2 pN, a piezo driven microscopic stage starts to move in the opposite direction, thus keeping the force constant. The feedback system had a range of 3.6 $\mu$m, enough to allow for determination of NcKin processivity under different external loads.

### 3.4.2.3 Sample preparation

Measurements were done in flow chambers, with a volume of about 5-15 $\mu$l (Fig. 3.15) (2.3.12). NcKin$_{hTail}$ constructs were bound via unspecific adsorption onto carboxylated latex beads with a diameter of 0.5 $\mu$m in. To prevent denaturation of the motor proteins on the bead surface casein as blocking reagent was present in all buffers at a concentration of 1 mg/ml. To ensure observation of single molecules, the loading density of kinesin on the beads had to be very low. Generally, kinesin concentration was adjusted in such a way that not more than 1/3 of the beads in a chamber showed movement along the microtubule. This corresponds to a small probability of about 6 % that one bead carries more than a single kinesin motor [Block et al., 1990; Jaud, 2003].

For visualization in the fluorescence microscope microtubules labelled with the fluorescent dye Cy5 were used in laser trap experiments. To immobilize them on coverslips, the flow chambers were pre-incubated with an anti-tubulin antibody that adhered unspecifically to glass surfaces. Since the feedback system worked only in one dimension, only microtubules that were correctly aligned could be used for the assays. Quickly flushing the flow chamber with the microtubule solution produced suitable alignment of the filaments.

## 3.4.3  Observation of single NcKin molecules in the optical trap

To detect the movement of single NcKin$_{hTail}$ and NcKin_stableNeck molecules, the kinesin constructs were bound at low density onto carboxylated latex beads (3.4.2.3) and placed above microtubules with the optical trap. Fig. 3.16 shows a typical trace of a single NcKin$_{hTail}$ molecule under 2 pN constant external load using the feedback system. In good agreement with previous observations [Crevel et al., 1999; Lakamper et al., 2003], several processive runs are visible in the feedback signal (black line) ranging from 0.5 to 1.4 $\mu$m. The red signal represents the backward force of the laser gradient corresponding to the bead position, which increases initially until reaching 2 pN and then stays constant due to feedback activation. When the motor dissociates from the microtubule, it is pulled back to its original position by the laser trap. Since the bead freely rotates within the laser beam, the motor situated on the bead will interact with the microtubule again, starting another processive run. The slope of the left flank of the feedback signal directly gives the velocity of the motor molecule.

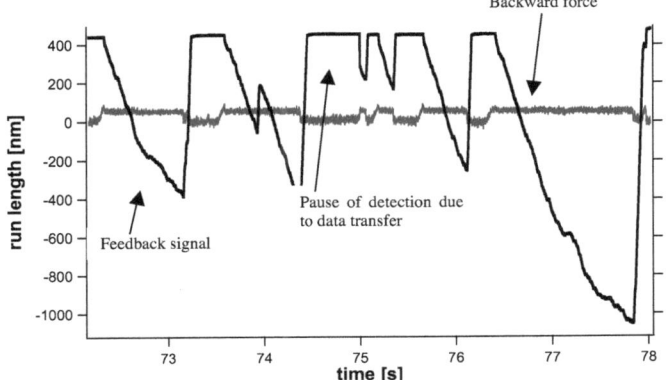

**Fig. 3.16: Processive runs of a single NcKin$_{hTail}$ molecule in the optical laser trap with activated feedback system under 2 pN external load.**
The displacement of the piezo-table (feedback signal, black line) represents the processive run of the motor molecule plotted versus the time, giving the length and the velocity (slope of the left flank). The red signal is the backward force onto the motor molecule that increases initially until reaching 2 pN and is subsequently held constant by the feedback. Every 5 seconds there is a pause of detection due to data transfer from the photo detector to the computer.

### 3.4.3.1 Stall forces of single NcKin molecules

Stall forces of single NcKin$_{hTail}$ and NcKin_stableNeck molecules were measured under saturating ATP concentrations (2 mM ATP) without using the feedback system.

Fig. 3.17 shows example traces of single NcKin-wt and stableNeck molecules. For both molecules, processive runs could be detected. As the motor moved the bead out of the centre of the laser beam, the backward force of the laser gradient increased until the maximum force of a single motor molecule was reached. At this point, the motor stopped moving, stalled for a certain time and finally dissociated from the microtubule and is pulled back by the laser gradient into the starting position.

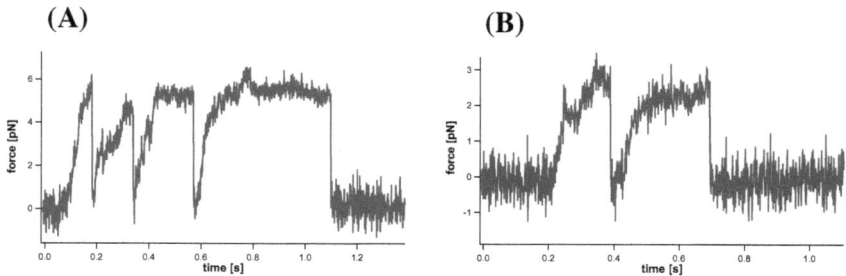

**Fig. 3.17: Runs of single NcKin molecules in the optical trap without feedback system.**
The bead displacement by a single motor molecule out of the centre of the laser beam is plotted versus the resulting backward force. By reaching the maximum force, the motor stalled for a certain time, dissociated from the microtubule and is pulled back into its original position, where it starts another processive run.
**(A):** $NcKin_{hTail}$; stall force in this example: ~ 5 pN
**(B):** NcKin_stableNeck; stall force in this example: ~ 3 pN

To determine the stall forces of single $NcKin_{hTail}$ and NcKin_stableNeck molecules, data from several experiments were pooled into a histogram (Fig. 3.18). Only those events were included, where the motor stalled for at least 1/20 second, since accidental dissociation during a processive run does not represent the maximum force of a motor. Moreover, it had to be assured that events originated from single molecules. Therefore, a very low kinesin concentration was chosen so that only 1/3 of the beads in a flow chamber displayed movement along the microtubule. In addition, runs that were caused by multiple motor molecules situated in the same small range on the bead surface, could be easily identified since those resulted in significantly higher stall forces of > 10 pN.

Stall forces determined with single Nckin-wt molecules lied in a range from 2.0 to 7.0 pN, with an average of 4.25 ± 0.44 pN for $NcKin_{hTail}$. Surprisingly, the NcKin_stableNeck construct displayed a significantly reduced stall force of 3.11 ± 0.35 pN, ~ 1.1 pN lower than the wild-type containing the "natural" neck domain. This indicates an important function of the fungal specific neck domain for the force production of single motor molecules.

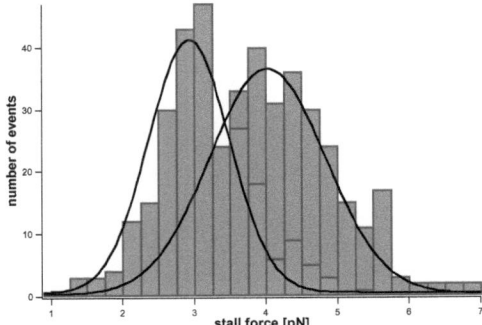

**Fig. 3.18: Histogram of the stall forces of single NcKin$_{hTail}$ (red) and NcKin_stableNeck (violet) molecules measured in an optical trap.**
Stall forces measured in several experiments were combined in the histogram. The black lines are Gaussian fits. The average stall forces are 4.25 ± 0.44 pN for NcKin-wt (N = 251) and 3.11 ± 0.35 pN for NcKin_stableNeck (N = 305).

### 3.4.3.2 Run length of single NcKin molecules under different external loads

To investigate the processivity of single kinesin constructs under different external loads, the average run length of NcKin$_{hTail}$ and NcKin_stableNeck was measured under 1 and 2 pN constant forces, using the activated feedback system. Data from several experiments were pooled into the histograms shown in Fig. 3.19. Similar to the stall force measurements only those events were included that originated from single motor molecules.

**(A)**

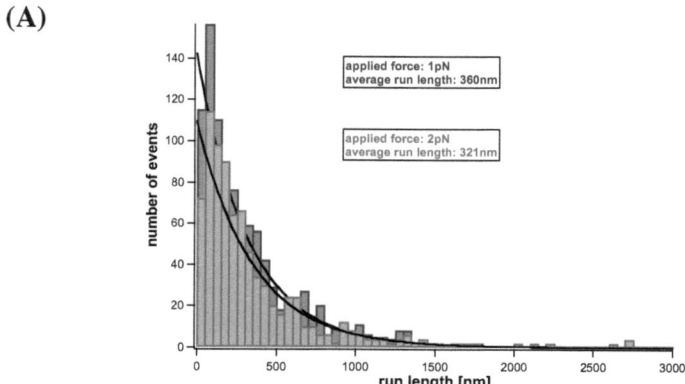

**(B)**

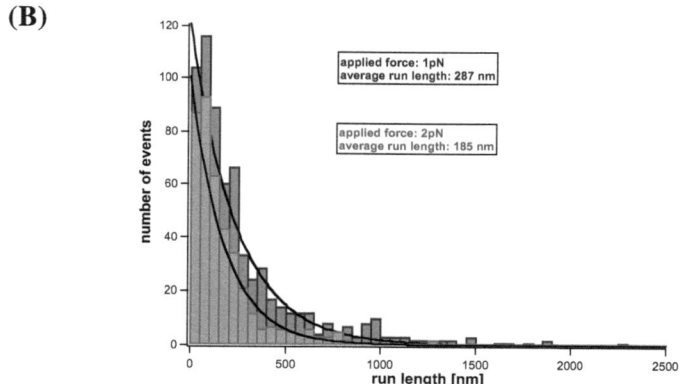

**Fig. 3.19: Run length distribution of NcKin$_{hTail}$ (A) and NcKin_stableNeck (B) under 1 and 2 pN force.**
Data from several experiments are combined in the histograms. The black lines are single exponential fits. The average values are given in the figure.

The average run length of the wild-type construct was 360 nm under 1 pN and 321 nm under 2 pN constant force, showing little influence on the processivity in this low-load regime. This was to be expected, since the stall force of NcKin$_{hTail}$ was determined as 4.25 pN (3.4.3.1), well above the external loads applied in this experiment.

A different situation emerged for the NcKin_stableNeck construct. As already observed in the stall force experiments, this mutant was able to move processively along the microtubule, indicating that the artificial neck domain does not affect the basic mechanical cycle of NcKin. However, the run length was significantly reduced with average values of 287 nm under 1 pN and 185 nm under 2 pN load. The significant decrease of the processivity (64 %) with increasing load reflects the poor force compliance of this mutant, as observed already in the stall force measurements.

### 3.4.3.3 Velocity of single kinesin molecules in the optical trap

As shown in Fig. 3.16, the slope of the left flanks of the feedback signal in the laser trap experiments gives the velocity of single motor molecules during a processive run under a constant force. Thus, the data of the run length measurements also could be used for velocity determination under 1 and 2 pN load. Data from several experiments are combined in the histograms shown in Fig. 3.20. Since the temperature could not be controlled in the laser trap experiments, the width of velocity distributions is relatively large.

At 1 pN force, the velocity of single wild-type molecules was lower than in multiple motor gliding assays (1.88 ± 0.19 $\mu$m/s compared to 2.7 $\mu$m/s, Tables 3.V), probably due to different ionic and temperature conditions in both assays. Under 2 pN external force, the velocity was reduced to 75% (1.40 ± 0.21 $\mu$m/s).

In contrast to the multiple motor assays, where the stableNeck construct produced only 50% of the wild-type velocity (Table 3.IX), single stableNeck molecules moved almost as fast as the wild-type in laser trap experiments (1.73 ± 0.23 versus 1.88 $\mu$m/s). Also similar to NcKin$_{hTail}$ constructs, the velocity of NcKin_stableNeck decreased to about 70% (1.20 ± 0.17 $\mu$m/s) under 2 pN external load.

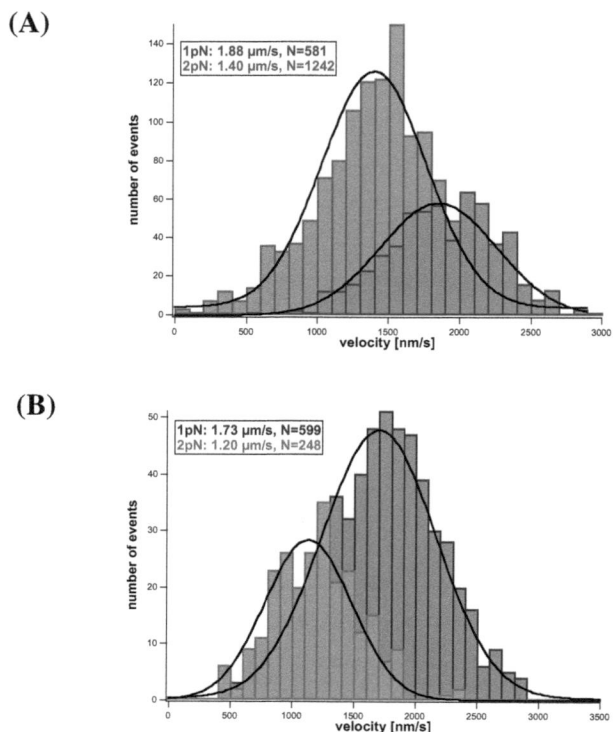

**Fig. 3.20: Histograms of the velocities of single NcKin$_{hTail}$ (A) and NcKin_stableNeck (B) molecules under constant forces in an optical trap.**
Velocity data from several experiments were combined in the histogram. The black lines are Gaussian fits. The average values are 1.88 ± 0.19 μm/s at 1 pN and 1.40 ± 0.21 μm/s at 2 pN force for NcKin$_{hTail}$. For NcKin_stableNeck, the average values are 1.73 ± 0.23 μm/s at 1 pN and 1.20 ± 0.17 μm/s at 2 pN force.

# 4 Discussion

Kinesins are microtubule-activated ATPases with a catalytic domain conserved within the entire superfamily [Vale et al., 1997; Woehlke et al., 2000]. The α-helical neck domain forms a coiled-coil in the dimer and works together with the heads in translating the energy derived from ATP hydrolysis into processive movement, but its precise role is poorly understood.

The filamentous fungus *N. crassa* possess an unusually fast kinesin-1 (NcKin) with a similar domain structure as its animal relatives [Steinberg et al., 1996]. In contrast to the motor domains, which exhibit a high degree of homology in fungal and animal conventional kinesins, fungal neck and hinge domains display a characteristic and strictly conserved sequence pattern, clearly different from animal neck domains.

To examine possible roles of the specific neck and hinge domains for the moving mechanism and/or the regulation of fungal kinesins, mutant NcKin constructs were generated, exchanging single, fungal-specific residues or the entire neck and hinge regions.

## 4.1 Complexity of neck dimerization and implications for the regulation of fungal kinesins

### 4.1.1 Functional importance of the conserved Tyr 362

The first part of this work addressed the functional importance of a highly conserved tyrosine at position 362 within the fungal neck domain. For this purpose, a mutagenesis approach was used, exchanging Y362 into lysine, the corresponding residue in the *Drosophila* conventional kinesin, cysteine or phenylalanine.

The motile and ATPase characteristics of $NcKin_{433}$-Y362K exhibited striking differences when compared to the wild-type construct $NcKin_{433}$: the $k_{cat}$ was enhanced three to four times, whereas the velocity in a multiple motor gliding assay was less than half of the wild-type (Table 3.I). This dramatic effect of the exchange of one amino acid residue in the neck resembles the consequences of the deletion of the entire neck domain of the *S. racemosum* kinesin [Grummt et al., 1998b]. As in the NcKin$_{433}$-Y362K point mutant, a 5-fold acceleration of the ATPase and a 2–3-fold reduction of the gliding velocity was observed. This amazing similarity hints at the prime importance of residue Y362 for proper neck functionality, in particular the effective coupling of ATP hydrolysis and movement of the motor. The basically identical basal ATP turnover rates of mutant and wild-type protein (Table 3.I) indicate that differences in motor functionality of the mutant are not intrinsic to the protein but are manifested in their interaction with the microtubule.

#### 4.1.1.1 Role of Tyr 362 for neck dimerization

At first sight, the defect caused by mutation of Y362 is due to a defective neck coiled-coil. The failure of NcKin$_{433}$ to dimerize when Y362 is exchanged shows that the interaction within the dimer is weakened in the mutant protein (Table 3.II). This is confirmed by the characterization of synthetic peptides comprising the neck domain of NcKin (aa 338-379) [Kallipolitou et al., 2001; Deluca et al., 2003]. The wild-type peptide clearly adopted a two-stranded α-helical coiled-coil conformation as indicated by circular dichroism only under acidic conditions (pH 3). Consistent with the coiled-coil formation, the ellipticity ratio $[\theta]_{222}/[\theta]_{208}$ was > 1 at pH 3, but decreases sigmoidally with increasing amounts of TFE, which is known to stabilize secondary structures, but disrupt tertiary structures [Su et al., 1994]. In contrast, the mutant peptide containing the Y362K displayed no coiled-coil formation under any pH conditions. Only at 5°C the mutant peptide was capable of assuming a coiled-coil conformation that, however, was less stable than the wild-type structure (Tm = 19.6°C versus 47.2°C for mutant and wild-type peptides, respectively). Thus, a salt bridge between Lys 362 (in g-position) and Glu 367 (in e-position) as promoted in the mutant does not stabilize the coiled-coil conformation [Deluca et al., 2003].

The effect of the Tyr substitution on the dimerization behaviour does not depend on whether a lysine or cysteine is introduced. The vicinity of the conserved aromatic residue F359 suggested that Y362 and F359 may form stacking interactions. However, the F359C mutant is still dimeric. On the other hand, the mutant Y362F, which does retain the aromatic side chain is monomeric (Table 3.II). Thus, stacking interactions are not relevant for coiled-coil stabilization, but the OH-group seems to be required. As Tyr is usually uncharged in proteins, stabilizing salt-bridges are unlikely. The exact mechanism, however, by which Y362 enhances the coiled-coil propensity remains unclear.

#### 4.1.1.2 Regulatory function of Tyr 362

A closer look at the effect of the Y362 exchanges reveals a more complex situation and suggests a second important role in addition to coiled-coil stabilization. The results on NcKin variants that are dimeric as wild-type protein but monomeric as mutant (NcKin$_{433}$) explain the dramatic effects of the point mutation as a consequence of a lack of head-head interaction, as required by the alternating head model of kinesin function [Hackney et al., 2003]. However, as indicated by the ATPase characteristics of the NcKin$_{383}$ constructs, stabilization of the neck structure is not the only function of Tyr 362. The ATP turnover of all Y362K/C mutants is accelerated up to 7-fold compared to their respective wild-type constructs. That Y362 not only stabilizes the α-helical neck coiled-coil is also supported by the Y362F mutation: it disrupts dimerization of NcKin$_{433}$ but does not lead to an accelerated ATPase rate (Tables 3.I and 3.II).

The simplest explanation for these observations is an inhibitory effect of the neck on the ATPase activity of the catalytic core, mediated by residue Y362 (Fig. 4.1). While the wild-type neck exerts an inhibitory effect on the ATPase by slowing down the steady-state rate from 260 /s to 24 /s, the mutant lacking the crucial tyrosine residue is unable to modulate the motor head, and thus shows a comparable rate as the neck-less deletion mutant. As indicated by comparison of Y362K/C and Y362F mutants, the aromatic ring of tyrosine may be sufficient to regulate the ATPase activity (Table 3.I), whereas the hydrophilic OH-group is necessary for proper coiled-coil inter-strand interaction (Table 3.II). Previous mapping studies of the NcKin motor revealed that monomeric constructs including the neck domain are 10-fold slower than the motor core without neck [Kallipolitou et al., 2001]. After proteolytic removal of residues from the neck domain the ATP turnover was accelerated. These data are in agreement with this work and strongly support the conclusion that the neck domain is able to repress the catalytic activity of the core motor domain.

Regardless of the exact structural basis of the inhibition, the mantADP release experiments indicate that Y362 slows down the microtubule-activated ADP release. In the mutant NcKin$_{383}$-Y362K, the ADP release is 6-7-fold faster than in the wild-type (Table 3.III). Formally, it cannot be excluded that other kinetic steps are accelerated, but as the mutant's mantADP release rate is much faster than the steady-state rate, it is thus sufficient to fully explain the effect.

In conclusion, these observations propose a dual function of the fungal neck domain via the conserved Tyr 362 (Fig. 4.1): first, a direct, regulatory action on the motor core and second, a more passive, indirect action on kinesin dimerization. For *Drosophila* kinesin it has been shown that the tail inhibition mechanism involves a slow-down of ADP release [Hackney et al., 2000]. One may therefore speculate that in the physiological case, Y362 mediates the regulatory influence of the tail region, which has been shown to down-regulate the ATPase activity [Coy et al., 1999; Friedman *et al.*, 1999; Seiler *et al.*, 2000]. Because monomeric constructs including the neck region of animal conventional kinesins do not exist, this regulatory influence of the neck domain remained previously undetected. Moreover, as Y362 is specifically conserved in fungal kinesins, regulation in animal conventional kinesins may involve different protein regions.

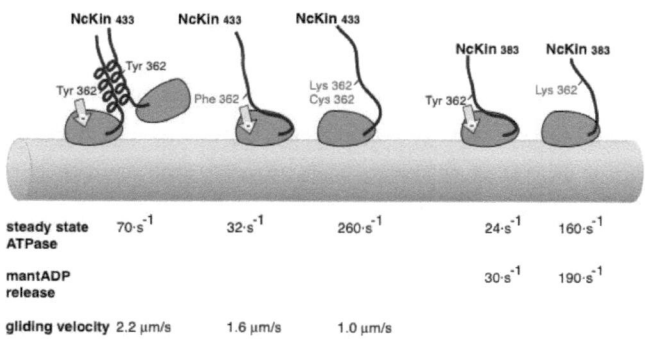

**Fig. 4.1: Schematic representation of the dual function of Y362.**
In the dimeric background (NcKin$_{433}$, left), the mutation leads to a disruption of the neck coiled-coil, resulting in a fast ATPase rate and slow gliding velocity. In the monomeric wild-type background (NcKin$_{383}$, right), Y362 mediates an interaction of an unfolded neck domain with the motor core, which represses the ATPase activity (yellow arrow). The K362 substitution is unable to regulate the core motor domain, which catalyses a fast ATP turnover comparable to the truncated NcKin$_{343}$ lacking the entire neck domain. The accelerated mantADP release rate in Y362K monomers indicated that the residue Y362 is involved in modulating the microtubule-activated ADP release.

### 4.1.2 Importance of the hinge region for the structural state of the NcKin neck domain

The specific sequence pattern of the NcKin neck and hinge domain is reflected in several unique properties. In contrast to animal kinesins, where the neck domain is sufficient to dimerize kinesin via α-helical coiled-coil conformation [Jiang et al., 1997b], fungal kinesins require the proximal part of the hinge domain to promote neck dimerization and thus to confer wild-type catalytic and motile properties [Kallipolitou et al., 2001].
To investigate the role of the specific hinge domain in fungal conventional kinesins in more detail, chimeric constructs were generated, replacing the NcKin hinge region by human kinesin sequences.

#### 4.1.2.1 Role of the conserved Trp 384 for NcKin neck dimerization

Although the amino acid sequences of all hinge regions diverge substantially, there is one strictly conserved residue, W384, near the N-terminus of the fungal hinge. To investigate the functional importance of this residue, two chimeric constructs were characterized, one retaining the conserved tryptophan (NcKin_384) and one lacking it (NcKin_379; Fig. 3.9). Only NcKin_384$_{hTail}$ displayed normal gliding and ATPase activities, whereas the NcKin_379$_{hTail}$ chimera was strongly affected (Table 3.V). Consistently, the C-terminally truncated NcKin$_{383}$ was monomeric and had low ATPase activity (Table 3.I and 3.II), whereas NcKin$_{391}$, only 7 residues longer but including W384 is a dimer and has high ATPase activity [Kallipolitou et al., 2001]. Moreover, a point mutant, NcKin$_{433}$W384F, exchanging W384 into phenylalanine, also failed to dimerize (G. Woehlke, personal communication). These data indicate that the conserved W384 plays an important role for the mechanism of fungal kinesin motors, most likely by promoting neck dimerization.
However, W384 is not sufficient to induce neck coiled-coil formation. In the 433-background NcKin_384 was shown to be monomeric although it includes W384 (Table 3.VI). The structural changes revealed by oligomerization studies are supported by functional assays. The ATPase activities of NcKin_384 and NcKin_379 are much lower in the 433 than in the hTail-background (Table 3.V). Since the unfolded neck domain was shown to inhibit the catalytic motor core via Y362, the reduced ATPase activity is a consequence of impaired dimerization.

#### 4.1.2.2 Evidence for intermediate structural states of the fungal neck domain

Since oligomerization studies in the 433-background do not necessarily reflect the situation in hTail constructs, an alternative method was used to gain information about the neck structure. To show spatial proximity, the endogenous P342 in the *a*-position of the first heptad repeat was replaced with cysteine to allow inter-subunit disulfide bond formation in coiled-coil structures [Zhou et al., 1993]. In previous studies with synthetic NcKin neck peptides containing P342C, crosslinking indeed stabilized the α-helical coiled-coil conformation [Deluca et al., 2003].

Evaluation of the disulfide bond formation on a non-reducing SDS-gel (Fig. 3.11) showed almost complete crosslinking of NcKin_379$_{hTail}$ and NcKin_384$_{hTail}$, similar to wild-type NcKin$_{htail}$ that contains P342C. This indicates close proximity of the two introduced cysteines. However, both chimeras were incapable of coiled-coil formation by themselves, as indicated by the monomeric state of the 433-versions (NcKin_379$_{433}$P342C and NcKin_384$_{433}$P342C, Table 3.VIII). Even more intriguing, the NcKin_379$_{hTail}$ chimera has low gliding and ATPase activities, as stated above, confirming the failure of the neck domain to adopt a functional folded state.

Why does a neck-neck crosslink still occur in this chimera? The observations hint at that the neck domain in the NcKin_379 construct is probably not completely disrupted but rather in a meta-stable state, allowing for crosslinking at P342C but unable to provide the stability of a fully functional neck. In NcKin_384, the structural state of the neck is shifted towards the dimeric state, probably due to W384, since it can be easily transformed into a functional, coiled-coil conformation, induced by the hTail-appendix. Thus, the presence of the hTail fosters coiled-coil formation, but only in the presence of certain key residues like W384 and Y362. The latter is an indispensable prerequisite for NcKin neck folding, since its mutation leads to complete disruption of the neck structure, indicated by impaired crosslink formation in the hTail background (Fig. 3.11). Taken together, the NcKin neck domain is not a static rod-like element, but can display dynamic switching between different intermediate structural states that depend on conserved key residues within the neck and hinge regions.

### 4.1.2.3 Integrity of the neck coiled-coil is required for NcKin motility

The hand-over-hand model for processive movement of kinesin-1 requires tight coupling of the two heads in a dimer to guarantee coordinated action [Romberg et al., 1998; Young et al., 1998; Hackney et al., 2003]. However, the unusually high velocity and less favoured coiled-coil formation of the NcKin neck domain (this work; [Kallipolitou et al., 2001] raised the question if the same mechano-chemical coupling mechanism also applies for fungal kinesins. As discussed above, the NcKin_379 construct displays slow microtubule gliding and ATP turnover as well as impaired dimerization in the 433-background. The unexpected crosslink of the NcKin_379$_{hTail}$ chimera raised the possibility to artificially link the two unfolded neck regions, providing a control if the observed defects of the mutant are really caused by uncoupled heads even in the dimeric hTail background.

Indeed, the gliding velocity of NcKin_379$_{hTail}$P342C under crosslinking conditions was as high as in the NcKin wild-type (Table 3.VII), showing that a tight connection of the neck domains is also a prerequisite for the movement of fast fungal kinesin motors. Surprisingly, the velocity remained high also under reducing conditions in the presence of 2–5 mM DTT and decreased only slightly in the presence of 10 mM DTT. Oligomerization studies clearly showed that the P342C mutation itself did not lead to the dimerization of the chimeric constructs under reducing conditions (Table 3.VIII). Therefore, it is very likely that the fast movement in the multiple motor assay is produced by a portion of molecules that are still crosslinked even in the presence of DTT. There are several arguments supporting this view:

(I) The neck/neck crosslink via C342 is very favourable since it appeared already under air oxidation (Fig. 3.11). For the gliding assay, the kinesin preparations had to be incubated on the coverslip several minutes to ensure adhesion on the glass surface (2.3.11.3). During that time, the molecules are exposed to air, so even in the presence of DTT the crosslink probability stays high.

(II) Only a small portion of crosslinked motor molecules is needed to produce high velocities in a multiple motor assay. Gliding experiments with mixed motor populations of fast NcKin wild-type and slow NcKin_379 chimera (with native P342) indicate that 10 % wild-type proteins generate fast velocities of ~ 2 μm/s. Therefore, as little as 5 to 10% of NcKin_379$_{hTail}$P342C molecules that remain cross-linked, can produce a gliding velocity of ~ 1.7 - 2 μm/s, even under reducing conditions (Table 3.VIII). According to Fig. 3.11, such a level of contamination is quite plausible.

(III) The ATP turnover under reducing conditions is low as expected, comparable to the parent construct with native P342 (Table 3.III). In contrast to the gliding assay the steady-state ATPase rate is nearly unaffected by incomplete cleavage of the disulfide bond in the protein preparation.

### 4.1.2.4 Stabilization of the neck coiled-coil in NcKin

It is a remarkable fact that in both animal and fungal conventional kinesins the neck sequence does not make for a perfectly matching coiled-coil. Particularly in the N-terminal part there are several non-ideal, hydrophilic or bulky residues in the *a* and *d* positions of the heptad repeat that appeared to destabilize α-helical coiled-coil conformation, for example Tyr344 and Glu347 in the human kinesin sequence [Tripet *et al.*, 1997; Thormahlen *et al.*, 1998] (Fig. 3.1). However, a new perspective on kinesin neck structure has emerged since a novel set of coiled-coil stabilizing motifs has been recognized for the human kinesin neck domain [Tripet *et al.*, 2002]. Helix-capping motifs that are formed not only by residues that lie within the neck domain but also in the adjacent neck-linker region, greatly stabilize the N-terminus of the α-helical neck coiled-coil and compensate the destabilizing effect of non-ideal residues. No specific helix-capping motifs have been found at the

C-terminal end of the neck, however, the strictly conserved W368 (W373 in the NcKin sequence) is thought to undergo stacking interactions that terminates the coiled-coil [Thormahlen et al., 1998]. Moreover, sequences similar to so-called trigger motifs that nucleate and/or stabilize α-helical coiled-coil conformations have been identified in the last two neck heptads [Kammerer et al., 1998; Steinmetz et al., 1998], that accordingly were found to be crucial for the coiled-coil stabilization in human kinesin neck regions [Tripet et al., 2002].

In the fungal sequence similar capping motifs as in the human case are present at the N-terminal end of the neck coiled-coil. According to Tripet et al. 2002, the hydroxyl group of S341 that resides in the neck-linker is a hydrogen donor for the backbone amid of E344, forming a so-called capping box. Another important residue in the human kinesin sequence is W340, which occupies a dual function in stabilizing the neck coiled-coil. First, it undergoes hydrophobic intra-chain interactions with L335, forming the hydrophobic staple motif, and second, the large indole ring is thought to interact extensively with the aliphatic residue L341 from the opposite chain. These two interactions, forming a so-called "molecular sandwich", are critical for the stabilization of the human neck coiled-coil [Tripet et al., 2002]. Since there is a leucine residue at the corresponding position (L345) in the fungal sequence, comprising a large aliphatic side chain but no aromatic ring, it is uncertain whether a similar molecular sandwich motif is used for coiled-coil stabilization in fungal kinesins, although hydrophobic staple interactions with L340 are likely.

As shown here, coiled-coil formation in the fungal neck is fostered by additional residues from the hinge domain, especially W384. This is surprising since this residue is clearly not part of the coiled-coil domain itself. NMR of rat kinesin hinge peptides shows the N-terminal portion of the hinge to consist of well-defined secondary structure elements [Seeberger et al., 2000], which are likely to occur also in fungal kinesins because of substantial sequence similarities at the neck-hinge junction. According to NMR data, the neck-hinge junction is highly flexible, allowing the C-terminal hinge to explore a large volume in space. Downstream of the flexible N-terminus of the hinge domain, there is an inflexible motif (E374-Q381 in rat kinesin). The residue Q381, corresponding to W384 in NcKin, is located at the tip of this region that conceivably could flip back to the central part of the neck helix (Fig. 4.2). The NMR structure model is consistent with the idea that W384 forms an inter-chain molecular sandwich motif with the C-terminal end of the neck and thus stabilizes the fungal neck coiled-coil.

### 4.1.3 Possible role of neck/hinge dynamics for the regulation of fungal kinesins

The preceding observations describe the complex and sophisticated dimerization behaviour of the fungal neck domain and present evidence for a dynamic switching between different structural states. Based on these findings the possibility of a functional role of neck/hinge dynamics for the regulation of fungal kinesins is conceivable. In several studies a regulatory influence of the tail domain on the motor domain has been described for animal and fungal kinesin-1. In the absence of the cargo, full-length kinesin adopts a compact conformation that juxtaposes the tail region and the head and neck domains of the motor [Stock *et al.*, 1999]. In this inhibited state, the globular tail domain lowers the ATPase activity and stepping rate via the conserved IAK motif, possibly by inhibiting microtubule-stimulated ADP release [Coy *et al.*, 1999; Hackney *et al.*, 2000; Seiler *et al.*, 2000]. Since the neck domain was found to be involved in this process, optimization of the coiled-coil sequences during evolution for motor regulation has been suggested [Friedman *et al.*, 1999; Kirchner *et al.*, 1999a].

The finding that a limited number of conserved key residues within the fungal neck and hinge sequences greatly influence the dynamic structure of the neck coiled-coil, as presented here, fits well into this picture. Since the hinge region can adopt many different orientations in space, it is conceivable that the tail region or a regulatory interaction partner of NcKin could interact with W384, inhibit the formation of a sandwich motif and prevent neck dimerization. In such a model the neck domain thus quickly switches between an "on" and an "off" state (Fig. 4.2).

In the "off" state, the kinesin molecule adopts a compact conformation. Interactions of the globular tail or other regulatory interaction partners with the neck domain capture W384 and thus prevent formation of a coiled-coil. As shown in this work, the other key residue within the fungal neck, Y362, inhibits the ATPase activity of the motor core when the neck is in the unfolded state (Fig. 4.1), thus preventing futile ATP hydrolysis of the motor in the switched-off situation. As shown previously, cargo binding releases the interaction of the tail domain with the N-terminal parts of the motor and thus induces the "on" state [Coy *et al.*, 1999; Seiler *et al.*, 2000]. The proximal part of the hinge region folds back, allowing W384 to stabilize the formation of a coiled-coil by undergoing hydrophobic stacking interactions. Dimerization of the neck domain releases inhibition of the motor core via Y362 and enables effective and processive movement of the motor protein. Confirming this model, the absence of W384 in the chimeric mutants does not completely unfold the neck domain, as in the case of Y362 mutations, but merely shifts the equilibrium towards a labile coiled-coil. This suggests the quick reformation of the active state of the neck, as expected for a regulatory mechanism. In a yeast two-hybrid screen a weak, presumably transient interaction of the tail domain (aa 724-928) with a truncated NcKin motor-neck construct including W384 (NcKin400) could be detected, but not with a truncated construct, lacking the proximal hinge (NcKin378) (L. Driller, personal communication). These preliminary findings further indicate the functional link between the hinge-based motor regulation as seen here and the tail-based regulation determined in previous studies.

Since the fungal key residues described here are not present in animal kinesin sequences it is uncertain whether a similar regulatory mechanism also applies to these motors. The importance of the conserved IAK motif for motor regulation in both subgroups suggests a common basic mechanism. Confirming the importance of the neck region for intra-molecular motor regulation, the replacement of the neck domain by an artificial sequence in human kinesin almost abolishes tail inhibition [Friedman *et al.*, 1999]. However, since the neck domain is sufficient for dimerization in animal kinesins, "long monomers" comprising the neck region do not exist, precluding the observation of neck inhibition as found here for NcKin. The presence of light chains indicates a more complex regulatory mechanism in animal kinesins, suggesting the involvement of additional interaction partners that trigger motor activation in higher regulation levels.

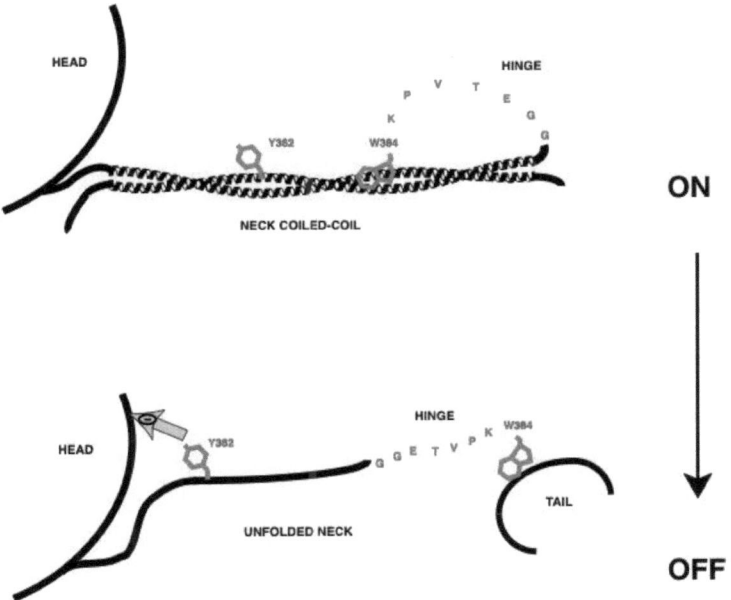

**Fig. 4.2: Model for the possible role of neck/hinge dynamics for the regulation of fungal kinesins.**
**ON state:** The neck is in the two-stranded coiled-coil state due to stabilization via W384 (conserved fungal sequences highlighted in orange). The hinge region is folded backwards to enable interaction of W384 with the neck domain, requiring large spatial freedom that is achieved by the highly flexible neck-hinge junction. Dimerization of the neck provides mechano-chemical coupling of the two heads and thus effective und processive movement, powered by fast ATP hydrolysis in the head domain.
**OFF state:** In the absence of cargo the tail domain in the full-length protein folds back (compact confomer) and captures W384. This greatly destabilizes the coiled-coil structure and thus transforms the neck domain to an unfolded state. The lack of head-head coupling stops processive movement of the motor; consequently, the ATPase activity is down-regulated due to the inhibitory interaction of Y362 with the motor core (Fig.4.1).

## 4.2 Role of the specific neck domain for NcKin motor mechanics

### 4.2.1 The motor core determines fast ATPase and gliding activity in fungal kinesins

All fungal kinesins studied so far move 4 to 5-fold faster than their animal relatives (2.2-2.7 μm/s versus 0.4-0.8 μm/s, respectively). Also the steady-state ATP turnover is faster; e.g. 60-80 /s for NcKin versus 20-50 /s for human and *Drosophila* kinesin [Jiang *et al.*, 1997b; Ma *et al.*, 1997a; Gilbert *et al.*, 1998]. Since comparison of the motor domains of animal and fungal kinesins reveals a high degree of sequence and structure conservation [Kirchner *et al.*, 1999b; Song *et al.*, 2001], the conserved and characteristic neck domain is a potential candidate for the determination of the fast ATPase and gliding activity. To address this issue, the kinetic and motile characteristics of mutant NcKin constructs lacking the specific neck or exchanging it with human kinesin sequences were investigated.

#### 4.2.1.1 Temperature dependence of NcKin and HsKin

Because of the close degree of similarity between fungal and animal kinesin-1, it has been suggested that the higher motile and catalytic activity of fungal motors may not be intrinsic to the protein but due to different temperature optima. This appeared to be reasonable since the optimal growth temperature of the fungus *N.crassa* is ~ 25°C, well below the human body temperature (37°C). However, the Arrhenius activation energies for ATP turnover and gliding activity turned out to be very similar for NcKin and HsKin (48-58 kJ/mol for HsKin and 37-49 kJ/mol for NcKin, Fig. 3.10) and agreed well with previous measurements ($\Delta E_A$ = 65 kJ/mol for full-length HsKin; [Bohm *et al.*, 2000]). At each temperature, ATP hydrolysis and microtubule gliding of the NcKin construct were faster than that of the HsKin construct (Table 3.IV). This clearly indicates that the higher motile and kinetic activity of the fungal motor is an intrinsic property and not a matter of temperature dependence. In addition, the parallel slopes of the Arrhenius plots suggest similar rate-limiting steps in the reaction cycles of both motor proteins, confirming a common basic mechanism. However, the higher "working temperature" of HsKin is reflected in a broader temperature activity interval, since microtubule gliding could be observed up to 40°C, whereas NcKin is already inactivated at temperatures > 28°C.

#### 4.2.1.2 The specific fungal neck domain can be replaced by other kinesin sequences.

To prevent any influences from C-terminal regions, the ATPase measurements at different temperatures were performed with truncated constructs, comprising only the motor core and the neck-linker. In agreement with previous observations, these "minimal motors" showed faster ATP turnover than dimeric constructs (Table 3.VIV) [Jiang *et al.*, 1997a; Ma *et al.*, 1997b; Moyer *et al.*, 1998; Kallipolitou *et al.*, 2001]. The extremely high ATPase rate of the NcKin head construct (~ 260 /s at 22°C) nicely correlates with the high gliding velocity of 2 $\mu$m/s obtained with NcKin dimers, however, these longer constructs exhibit a much lower ATPase activity (~ 70 /s at 22°C). The reason for this discrepancy is not known since an inhibitory influence of the tail domain on the ATP turnover was ruled out by using truncated hTail-constructs in the motility assays. Nonetheless, the high ATPase activity of NcKin head constructs implies that the catalytic motor core contains determinants for the fast gliding velocity of NcKin. Still, the characteristic and conserved neck sequences may be important for transforming the fast ATPase into fast movement.

To address the functional role of the fungal neck domain in the generation of fast movement it was replaced by human kinesin sequences with and without concomitant exchange of the hinge domain (NcKin_340 and NcKin_340-79, Fig. 3.9). Surprisingly, both chimeric constructs exhibited fast movement with a similar gliding velocity as the wild-type (Table 3.V). The NcKin_340 also displayed fast ATP turnover, whereas the NcKin_340-79 chimera, retaining the NcKin hinge domain, showed a slight reduction in catalytic activity as both hTail and 433-version. However, since this construct moved nearly as fast as the wild-type motor and was unaffected in its dimerization behaviour (Table 3.VI), the decreased ATP turnover was not considered to be a true phenotype of the domain exchange.

These findings indicate that the specific fungal neck domain is not required to transform fast ATP hydrolysis into fast movement, but that the high ATPase and gliding activity of fungal kinesins is exclusively determined by the motor core.

### 4.2.2 Mechanic properties of an artifical coiled-coil in the fungal neck domain

#### 4.2.2.1 Rationale for the design of the NcKin_stableNeck mutant

The results stated in the previous section suggest that the neck domain in fungal kinesins plays a passive structural role, providing a tight connection between the two heads in a dimer, with the conserved and specific residues having no important function for motor mechanics. This was tested by replacing the neck domain with an artificial coiled-coil sequence, bearing no similarities with the natural fungal neck domain. For the insertion the so-called EIEALKA sequence was chosen, designed by Su et al. (1994) to maximize hydrophobic and ionic intra- and interchain interactions. Characterization of a 33-aa peptide of this sequence revealed a remarkable stable coiled-coil conformation with a GdnHCl denaturation midpoint of > 7 M and a melting temperature of ~ 140 °C [Su *et al.*, 1994]. Synthetic peptides comprising the natural NcKin neck sequence exhibited much lower coiled-coil stability with a melting temperature of 47 °C. Thus, replacement of the neck domain with 5 heptads of the EIEALKA sequence had two effects: First, all of the specific fungal residues were exchanged by the artificial sequence. Only 7 leucine, lysine and isoleucine residues remained unchanged since they match the basic sequence requirements for a coiled-coil conformation and are also present in other kinesin neck sequences (Fig. 3.13). Second, the artificial neck domain provides a much more stable coiled-coil structure than the natural fungal neck. Thermodynamic studies revealed a $\Delta G$ for denaturation of > 80 kJ/mol [Tripet *et al.*, 1997]. Since the free energy derived from ATP hydrolysis is about 40–50 kJ/mol, unfolding of the coiled-coil during the reaction cycle is very unlikely [Romberg *et al.*, 1998].

#### 4.2.2.2 The artificial neck satisfies basic requirements for NcKin motility

NcKin dimerization by the artificial neck domain was confirmed by determining the oligomerization state of the NcKin_stableNeck$_{433}$ mutant in gel filtration and sucrose density centrifugation, and, as expected, the mutant clearly appeared as a dimer (73 kD versus 47.9). This finding agrees well with the stable coiled-coil conformation of peptides comprising the EIEALKA sequence [Su *et al.*, 1994; Tripet *et al.*, 1997] and the single molecule behaviour of the NcKin_stableNeck construct as described below.

Surprisingly, the velocity of the NcKin_stableNeck was significantly reduced in the multiple motor gliding assay to about 50 % of the wild-type value (Table 3.IX), whereas the ATPase activity was almost not affected. This indicates a less effective coupling of ATP hydrolysis and movement due to the artificial neck sequence. In a previous study the replacement of the human kinesin neck coiled-coil with 4 heptads of the EIEALKA sequence also resulted in a decreased velocity in a multiple motor gliding assay, but not in a single molecule motility assay observing individual kinesin molecules moving along an axoneme [Vale *et al.*, 1996a; Romberg *et al.*, 1998]. To characterize the effect of the artificial neck sequence on NcKin motor-mechanics in more detail, single molecule assays were performed in an optical laser trap. Under different backward forces single NcKin-wt and NcKin_stableNeck molecules both showed processive movement (Fig. 3.17), indicating that the artificial coiled-coil sequence provides a functional head-head connection to ensure coordinated movement. Importantly, the velocity of individual molecules in the laser trap assay was very similar for NcKin-wt and NcKin_stableNeck under low load conditions (1 pN; 2.16 µm/s and 1.88 µm/s, respectively), consistent with previous findings [Romberg et al., 1998]. Taken together, these observations confirm that the specific NcKin neck domain is not required for fast movement and can be replaced by both, other kinesin necks or artificial coiled-coil sequences.

### 4.2.2.3 Neck unwinding is not essential for NcKin stepping

In the crystal structure of a kinesin dimer the two motor domains cannot bind simultaneously to the microtubule [Kozielski et al., 1997], as required for the hand-over-hand model of kinesin processivity [Hackney, 1994a; Hirose et al., 1996]. This and the presence of unfavourable residues in the first half of the neck coiled-coil had led to the proposal that a segment of the neck domain may unfold transiently during the reaction cycle, allowing the two heads to span the 8 nm between two consecutive binding microtubule binding sites [Tripet et al., 1997; Hoenger et al., 1998; Mandelkow et al., 1998; Hoenger et al., 2000]. For animal kinesin this was rendered unlikely by the study mentioned above, that replaced the neck domain by the stable EIEALKA sequence [Romberg et al., 1998] and by the introduction of a crosslink at the N-terminus of the neck coiled-coil [Tomishige et al., 2000]. Both mutations did not affect the velocity and reduced the processivity of single motor proteins relatively mildly (by 16-50 %). However, for fungal kinesins a (partial) unwinding of the neck domain could not be ruled out completely because of their unusually high velocity and the less favoured neck dimerization (this work and [Kallipolitou et al., 2001]).

The findings presented here strongly argue against a requirement for neck unwinding during NcKin movement since single stableNeck molecules move processively and produce similar velocities as the wild-type. Despite the large unfolding energy of > 80 kJ/mol, it cannot be totally excluded that unwinding of the artificial sequence still occurs due to favourable entropic influences in the protein environment. However, the introduction of a crosslink at the N-terminus of the neck region in NcKin also did not affect ATPase and gliding velocity in a multiple motor assay [Hahlen, 2004a], confirming that neck dissociation is not a critical element for basic motor mechanism. In fact, constructs with a neck stabilized by a crosslink support fast movement and high ATPase activity better than constructs with a floppy neck, as shown with the crosslinked NcKin_379-P342C construct (4.1.2.3). Thus, together these studies clearly show that a stable neck coiled-coil is compatible with fast movement.

#### 4.2.2.4 Single molecule behaviour of NcKin wild-type

In laser trap experiments single NcKin wild-type molecules exhibited the typical behaviour of processive motors, in good agreement with previous findings [Crevel *et al.*, 1999; Lakamper *et al.*, 2003]. The beads made repeated excursions from the trap centre (Fig. 3.16/17), reaching and holding stall force, eventually detaching from the microtubule and being pulled back into the centre. The mean stall force of 4.25 pN also agreed with the results of other groups for NcKin (~ 5.0 pN, [Crevel *et al.*, 1999]), being slightly lower than the maximum force of human kinesin under similar conditions [Svoboda *et al.*, 1994; Meyhofer *et al.*, 1995; Visscher *et al.*, 1999]. Under the low loads applied in the experiments (1 pN), single molecule velocity was high (1.88 ± 0.39 µm/s), but still considerably lower than in multiple motor gliding assays (2.7 µm/s), probably due to different temperature and ionic conditions.

Interestingly, the run length of NcKin wild-type under 1 pN external load was substantially smaller (0.36 µm) than previously reported from single molecule motility experiments (1.75 µm, [Lakamper *et al.*, 2003]). This difference cannot be attributed to the zero-load conditions in single molecule fluorescence assays, since the backward force in the laser trap experiments of 1 pN lies well below the maximum force of NcKin, as stated above. In fact, single NcKin motors exhibited an unusual running behaviour in the laser trap, since they frequently dissociate during processive runs and rebind to the microtubule until they detached completely and were pulled back into the trap centre. Usually, rebinding events are regarded as evidence for many motors acting together on one bead. However, this appeared unlikely in these experiments because of statistical considerations (1/3 of the beads displaying movement along the microtubule corresponds to only 6 % probability of more than a single motor attached to one bead). Moreover, in practice multiple motor events could be easily identified since they produce a significantly higher stall force. Because consecutive backward stepping was never observed for conventional kinesin motors, even under over-stall loads [Coppin *et al.*, 1997], backsliding of more than 16 nm before rebinding to the microtubule was regarded as a new run. Thus, processive runs that appeared to be very long at first sight, turned out to be a series of several shorter runs. However, short dissociation events in the range of 20-100 nm cannot be resolved in the fluorescence microscope, thus, in single molecule motility assays a series of events were detected as one long, processive run, resulting in a much higher mean value.

#### 4.2.2.5 Single molecule behaviour of NcKin_stableNeck

Although the nucleotide-dependent unwinding of the neck coiled-coil appears not to be essential for the processive cycle of fungal kinesins, the neck domain obviously displays more than a passive structural connection between two motor domains, since single molecule experiments revealed interesting, more subtle effects with the NcKin_stableNeck mutant containing the stiff coiled-coil neck sequence.

Interestingly, stall force measurements in the fixed optical trap at saturating ATP concentrations revealed a significant decrease in force production of the NcKin_stableNeck construct compared to the wild-type (3.11 pN versus 4.25 pN, respectively). The reduced force compliance of the stableNeck mutant is also reflected in the processivity of single motors under different constant loads. With 1 pN, the NcKin_stableNeck construct exhibited a slightly reduced run length (by 23 %) compared to the wild-type, whereas under 2 pN load the processivity decreased substantially to ~ 60 % of the 1 pN value and ~ 50 % of the corresponding wild-type run length (Fig. 3.19). These findings are consistent with previous reports where the processivity of single kinesin motors strongly decreased under high external loads approaching the stall force [Coppin et al., 1997; Visscher et al., 1999]. Since the wild-type NcKin motor exhibited a significantly higher maximum force (4.25 pN), the run length was nearly unaffected in this low load-regime.

In contrast, the velocity of single molecules decreased considerably under 2 pN load to ~ 70-75% of the 1 pN-value in both, wild-type and stableNeck mutant. Thus, processivity and velocity are affected to different extents and the velocity is more force-sensitive also in the wild-type. It will be interesting in the future to study this behaviour in more detail and to find out whether it has any functional significance for fungal motors.

It is noteworthy that single molecule observations in the laser trap additionally revealed a more qualitative difference between the NcKin_stableNeck and wild-type construct. The mutant motor moved less smoothly than the wild-type, with frequent pauses during processive runs, even before reaching the stall force. This was not due to protein inactivation during the measurements, since the same molecule continued moving with similar velocities before and after the pause. The exact reason for this behaviour is not clear from these experiments, however, it further indicates the importance of some flexibility within the neck domain for NcKin motor mechanics.

The distinct features of single NcKin_stableNeck motors described above explain the discrepancy in the velocities obtained from multiple motor gliding assays and single molecule bead assays with this construct (1.88 $\mu$m/s versus 1.3 $\mu$m/s). In the multiple motor assay thousands of kinesin molecules are acting together on one microtubule (app. motor density: 28000 molecules per 1 $\mu m^2$). If one motor molecule pauses for a short time interval, as frequently observed with single stableNeck motors in the laser trap experiments, the other molecules are probably significantly impeded due to their reduced force compliance. This results in a lower net-velocity in multiple motor assays compared to single motors. Consistently, the velocity in multiple motor gliding assays increases with decreasing motor densities on the coverslip (Melanie Reisinger, personal communication).

### 4.2.2.6 Fine-tuning of NcKin motility by the neck domain

What are the conclusions for the physiological role of the kinesin neck domain from these single molecule observations?

Previously, fine-tuning of kinesin processivity by the neck domain has been suggested, mediated by electrostatic interactions between the positively charged neck coiled-coil and the negatively charged COOH terminus of the microtubule [Thorn et al., 2000; Wang et al., 2000]. This hypothesis is based on mutant human kinesin constructs that exhibited a gain in processivity by increasing the positive charge in the neck domain, whereas the introduction of negative charges resulted in decreased processivity. However, there is no direct evidence for such an interaction, since NcKin was clearly shown to be processive in laser trap experiments (this work and [Crevel et al., 1999]) and single molecule motility assays [Lakamper et al., 2003], although it carries no charge in the neck domain. Moreover, no effect in steady-state ATPase activity or microtubule gliding was obtained upon replacing the neutral fungal with the positively charged human neck domain (net charge: +4). In fact, insertion of 5 heptads of the EIEALKA sequence - comprising an extremely negative net charge of –7 – in fungal or the human kinesins would be expected to almost abolish processive movement, but instead resulted in a relatively mild effect.

The findings presented here do indicate a fine-tuning of kinesin motility by the neck domain, but based on a different mechanism. Importantly, the exchange of the natural fungal neck domain by the human kinesin neck leaves kinetics and motility unaffected, whereas the insertion of a stable neck coiled-coil significantly reduces force production and processivity. Similarly, reducing the flexibility of the neck domain in human kinesin, either by exchange with the EIEALKA sequence or by crosslinking, also resulted in a decreased processivity [Romberg et al., 1998; Tomishige et al., 2000]. It is therefore conceivable that the observed phenotype of the stableNeck mutant is not caused by the lack of specific fungal sequences but is due to decreased flexibility of the neck domain.

Since the mutation predominantly affects force production, a load-dependent step in the reaction cycle is presumably facilitated by a certain extent of structural flexibility in the neck domain. In the mechano-chemical cycle of kinesin proposed by [Schnitzer et al., 2000] based on measurements in a force clamp [Visscher et al., 1999], one 8 nm step of kinesin is accomplished by two sequential 4 nm steps (Fig. 4.3). The existence of distinct sub-steps within kinesin movement is controversial but has been observed twice [Coppin *et al.*, 1996; Nishiyama *et al.*, 2001]. The first of the two rapid 4-nm components is triggered by effective ATP binding to the attached head (1), followed by a subsequent load-dependent isomerization step (presumably neck-linker docking on the motor domain; [Rice *et al.*, 1999; Vale *et al.*, 2000]). The motor thus reaches an intermediate state (2), where the free head undergoes rapid 4-nm fluctuations forth and back along the microtubule lattice, caused by transient interactions of its neck-linker with the motor core [Rice *et al.*, 2003]. Eventually, one of these fluctuations leads to the binding of the free head to the next microtubule binding site and subsequently release of ADP, inducing the second 4-nm advance (3). External load has been demonstrated to reduce ATP affinity to the attached head and furthermore increase the time the motor spends in the intermediate state preceding the second 4-nm substep. During that stage, it is conceivable that some flexibility in the neck coiled-coil facilitates the diffusional search of the free head, but is not absolutely required to allow for reaching the next binding site, thus providing an optimization mechanism for kinesin processivity. In the stableNeck mutant, fluctuational freedom of the free head is probably decreased by the stiff coiled-coil domain, which becomes especially evident under load, thus explaining the reduced force production of this mutant. From these experiments, it cannot be decided if structural flexibility is provided by a slight unwinding of the neck coiled-coil or just by the structural features of the specific sequences in the neck domains, which may form a less stiff coiled-coil conformation than the artificial model sequence.

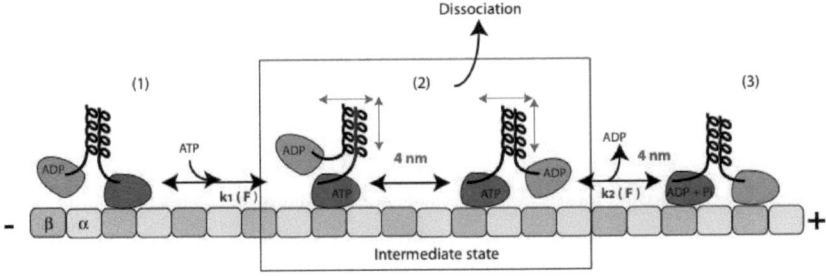

**Fig. 4.3: Fine-tuning of kinesin-1 processivity by the neck domain.**
In the proposed mechano-chemical cycle of kinesin-1 according to [Schnitzer *et al.*, 2000], the 8 nm step is composed of two 4 nm advances. Productive ATP binding to the attached head in state (1) triggers a load-dependent conformational change, transferring the motor to an intermediate state (2). Here, the unbound head undergoes rapid 4 nm-fluctuations forth and back along the microtubule lattice, until it eventually reaches the next binding site and releases its ADP, inducing the second, load-dependent 4 nm advance. External load reduces ATP affinity to the attached head and furthermore increases the time the motor spends in the intermediate state preceding the second 4-nm substep. Vertical or horizontal flexibility within the neck coiled-coil probably facilitates the diffusional search of the second head, thus shortening the intermediate state. Vice versa, increasing the neck stiffness as in the stableNeck mutant extends the intermediate state, reducing force compliance and processivity of the motor, since dissociation readily occurs in this state, where only one of the heads is attached to the microtubule.

## 4.3   Conclusions and future prospects

The detailed analysis of mutant NcKin neck constructs revealed a variety of functional roles of the specific neck domain in fast fungal kinesins. First of all the neck domain is, of course, responsible for motor dimerization as a basic requirement to ensure coordinated action of the two head domains and to hold the motor on the track during processive movement. For animal kinesins this has been studied extensively [Berliner *et al.*, 1995; Romberg *et al.*, 1998; Young *et al.*, 1998; Hackney *et al.*, 2003] and has been confirmed also for fungal kinesins in this work and in previous studies [Grummt *et al.*, 1998b; Kallipolitou *et al.*, 2001]. However, as shown by the chimeric constructs, NcKin dimerization is not limited to the neck domain but involves additional conserved residues from the adjacent hinge region, thus combining neck and hinge domains into one functional unit. This probably provides a sophisticated regulation mechanism in fungal kinesins, switching the neck domain from an unfolded conformation in the inactivate state of the motor to a correctly folded coiled-coil in the activated state. This dimerization-based regulation mechanism is mediated by several conserved key residues that may interact with the regulatory tail region of the motor or other interaction partners. Finally, the structural features of the neck domain are also important for fine-tuning the processive movement of conventional kinesins. Although transient melting of the neck coiled-coil does not occur during NcKin motility, a limited flexibility in the neck structure optimizes progression of the motor, especially under load. Thus, the specific and conserved sequence in the fungal neck domain has evolved to combine the different demands for optimal kinesin functionality like tight connection of the motor heads, flexible tether and regulatory unit in one short domain.

Certainly there are still lots of open questions. The proposed regulation mechanism of fungal kinesins via neck dimerization has to be confirmed experimentally and characterized in more detail. Does the tail domain really interact with W384, or are other regulatory interaction partners involved? How does cargo binding induce motor activation? Also the inhibition of the ATP turnover in the motor core by Y362 is still poorly understood. Which region in the motor domain is affected? *In vivo* characterization of mutant constructs as well as biochemical data addressing the influence of the tail region on ATPase kinetics and neck structure would help to clarify these issues. However, most importantly, structural data from the neck and hinge domains are needed to investigate how these regions are interacting with each other and with the motor core. Crystal structures of kinesin dimers including the hinge as well as structural characterization of neck-hinge peptides will provide direct information about the functional interplay of these regions.

Furthermore, the detection of discrete steps in single molecule experiments under low ATP conditions would be very useful to learn more about NcKin mechanics and the influence of the neck domain. Why does the stableNeck mutant pause so frequently during processive runs? Are there backward steps? What are the rate-limiting steps in NcKin and animal kinesin reaction cycles? Are there differences that could account for the increased velocity in fungal kinesins? How is the coupling efficiency in fast motors? In solving these problems, fungal kinesins as "natural mutants" would provide very useful model systems to improve understanding of conventional kinesin motors.

# 5    Summary

Fungal kinesin-1 (conventional kinesin) motor proteins like NcKin from the filamentous fungus *Neurospora crassa* exhibit a similar domain structure as their animal relatives: The N-terminal motor domains or "heads" with the ATP and microtubule binding sites are followed by the α-helical neck domains that form a coiled-coil in the dimer and work together with the heads to generate force and motility. The flexible hinge domain connects the N-terminal part of the motor with the C-terminal stalk and tail domains that are involved in cargo binding and regulation.

The most striking property of fungal kinesins is their high velocity: With 2 - 2.8 μm/s they move 3-4 times faster than their animal counterparts (0.5 - 0.8 μm/s). In contrast to their motor domains that exhibit great sequence similarities, fungal neck domains display a characteristic and strictly conserved sequence pattern, clearly different from animal neck domains. The hinge regions show poor sequence conservation in all kinesin-1 motors. There is, however, one strictly conserved Trp residue in the proximal part of the hinge, not present in animal sequences. Goal of the present work was to elucidate the functional roles of the specific neck and hinge domains for mechano-chemistry, dimerization and regulation of fast fungal kinesins. To address this issue, mutant constructs were generated, exchanging single, fungal specific residues or the entire neck and hinge regions with human kinesin or artificial sequences. The constructs were characterized in terms of oligomerization, motility and kinetic properties.

The exchange of the fungal-specific Tyr362 into Lys, Cys or Phe led to a loss of dimerization. Whereas the Phe substitution had only a structural effect, the Lys and Cys replacements also resulted in dramatic kinetic changes. The steady-state ATPase was accelerated 4- to 7-fold, probably due to a faster microtubule-stimulated ADP release rate. These data suggest an inhibitory effect of the fungal neck domain on the motor core, mediated by direct interaction of the aromatic ring of Tyr362 with the head, whereas its OH-group is essential for dimerization.

Previous studies on truncated NcKin constructs have shown that in contrast to animal kinesins the fungal neck domain is not sufficient for dimerization but needs additional residues from the hinge region [Kallipolitou *et al.*, 2001]. To investigate the structural requirements of NcKin dimerization, the fungal hinge was replaced by human kinesin sequences. The resulting chimera exhibited slow ATPase activity and microtubule gliding velocity and failed to dimerize. The motor function was restored by an intra-chain crosslink of the two neck domains in the first position, revealing that a tight connection of the neck domains is required for NcKin motility. Interestingly, a slightly longer chimera that retained the fungal-specific Trp384 in the proximal hinge, showed wild-type dimerization, ATP turnover and microtubule gliding velocity. In the absence of a stalk domain, however, W384 was not sufficient to induce dimerization, indicating complex dimerization behaviour of the neck that involves several intermediate structural states. The equilibrium between these states is highly dependent on key residues within the neck (Tyr362) and adjacent domains (Trp384). Based on these findings a regulatory mechanism for fungal kinesins is proposed, where the N-terminal part of the hinge folds back and Trp384 promotes neck dimerization by hydrophobic stacking interactions.

Despite their important function for motor regulation, the specific neck and hinge regions are not critical for the basic mechanism of fast fungal kinesins, since replacement with the human kinesin counterpart did not reduce the high microtubule gliding and ATPase activities of the wild-type. When the neck coiled-coil was substituted with an artificial, stable coiled-coil, however, the multiple motor gliding velocity was intermediate between animal and fungal kinesins. In contrast, single-molecule bead assays in an optical laser trap microscope showed that individual molecules of this mutant were able to produce maximum speeds similar to the wild-type. However, the mutant was heavily compromised in its ability to produce force. Even under low loads, the run lengths were significantly reduced. Together, these data suggest that the conserved sequence motifs in the neck domain have evolved to fine-tune kinesin-1 processivity by combining tight structural connection of the heads with a flexible tether that facilitates the diffusive search of the free head to the next binding site.

# 6 References

Andrews, P. (1970). Estimation of molecular size and molecular weights of biological components by gel filtration. Methods Biochem Anal *18*, 1-53.

Asbury, C. L., Fehr, A. N., and Block, S. M. (2003). Kinesin moves by an asymmetric hand-over-hand mechanism. Science *302*, 2130-2134.

Ashkin, A., Dziedzic, J. M., and Yamane, T. (1987). Optical trapping and manipulation of single cells using infrared laser beams. Nature *330*, 769-771.

Ashkin, A., Schutze, K., Dziedzic, J. M., Euteneuer, U., and Schliwa, M. (1990). Force generation of organelle transport measured in vivo by an infrared laser trap [see comments]. Nature *348*, 346-348 issn: 0028-0836.

Berliner, E., Young, E. C., Anderson, K., Mahtani, H. K., and Gelles, J. (1995). Failure of a single-headed kinesin to track parallel to microtubule protofilaments [see comments]. Nature *373*, 718-721.

Block, S. M., Goldstein, L. S., and Schnapp, B. J. (1990). Bead movement by single kinesin molecules studied with optical tweezers [see comments]. Nature *348*, 348-352.

Bohm, K. J., Stracke, R., Baum, M., Zieren, M., and Unger, E. (2000). Effect of temperature on kinesin-driven microtubule gliding and kinesin ATPase activity. FEBS Lett *466*, 59-62.

Bradford, M. M. (1976). A rapid and sensitive method for the quantitation of microgram quantities of protein utilizing the principle of protein-dye binding. Anal Biochem *72*, 248-254.

Brady, S. T. (1985). A novel brain ATPase with properties expected for the fast axonal transport motor. Nature *317*, 73-75.

Cantor, C. R., and Schimmel, P. R. (1980). Techniques for the Study of Biological Structure and Function. In Biophysical Chemistry (San Francisco, W. H. Freeman).

Case, R. B., Rice, S., Hart, C. L., Ly, B., and Vale, R. D. (2000). Role of the kinesin neck linker and catalytic core in microtubule-based motility [see comments]. Curr Biol *10*, 157-160.

Coppin, C. M., Finer, J. T., Spudich, J. A., and Vale, R. D. (1996). Detection of sub-8-nm movements of kinesin by high-resolution optical-trap microscopy. Proc Natl Acad Sci U S A *93*, 1913-1917.

Coppin, C. M., Pierce, D. W., Hsu, L., and Vale, R. D. (1997). The load dependence of kinesin's mechanical cycle. Proc-Natl-Acad-Sci-U-S-A *94*, 8539-8544 issn: 0027-8424.

Correia, J. J., Gilbert, S. P., Moyer, M. L., and Johnson, K. A. (1995). Sedimentation studies on the kinesin motor domain constructs K401, K366, and K341. Biochemistry *34*, 4898-4907.

Coy, D. L., Hancock, W. O., Wagenbach, M., and Howard, J. (1999). Kinesin's tail domain is an inhibitory regulator of the motor domain [see comments]. Nat-Cell-Biol *1*, 288-292 issn: 1465-7392.

Crevel, I., Carter, N., Schliwa, M., and Cross, R. (1999). Coupled chemical and mechanical reaction steps in a processive Neurospora kinesin. Embo J *18*, 5863-5872.

Crevel, I. M., Lockhart, A., and Cross, R. A. (1996). Weak and strong states of kinesin and ncd. J Mol Biol *257*, 66-76.

Cross, R. A. (2004). Molecular motors: kinesin's interesting limp. Curr Biol *14*, R158-159.

Dagenbach, E. M., and Endow, S. A. (2004). A new kinesin tree. J Cell Sci *117*, 3-7.

de Cuevas, M., Tao, T., and Goldstein, L. S. (1992). Evidence that the stalk of Drosophila kinesin heavy chain is an alpha-helical coiled coil. J Cell Biol *116*, 957-965.

Deluca, D., Woehlke, G., and Moroder, L. (2003). Synthesis and conformational characterization of peptides related to the neck domain of a fungal kinesin. J Pept Sci *9*, 203-211.

DeLuca, J. G., Newton, C. N., Himes, R. H., Jordan, M. A., and Wilson, L. (2001). Purification and characterization of native conventional kinesin, HSET, and CENP-E from mitotic hela cells. J Biol Chem *276*, 28014-28021.

Desai, A., Verma, S., Mitchison, T. J., and Walczak, C. E. (1999). Kin I kinesins are microtubule-destabilizing enzymes. Cell *96*, 69-78 issn: 0092-8674.

Diefenbach, R. J., Mackay, J. P., Armati, P. J., and Cunningham, A. L. (1998). The C-terminal region of the stalk domain of ubiquitous human kinesin heavy chain contains the binding site for kinesin light chain. Biochemistry *37*, 16663-16670 issn: 10006-12960.

Endow, S. A. (2003). Kinesin motors as molecular machines. Bioessays *25*, 1212-1219.

Endow, S. A., and Higuchi, H. (2000). A mutant of the motor protein kinesin that moves in both directions on microtubules. Nature *406*, 913-916.

Friedman, D. S., and Vale, R. D. (1999). Single-molecule analysis of kinesin motility reveals regulation by the cargo-binding tail domain [see comments]. Nat-Cell-Biol *1*, 293-297 issn: 1465-7392.

Funatsu, T., Harada, Y., Higuchi, H., Tokunaga, M., Saito, K., Ishii, Y., Vale, R. D., and Yanagida, T. (1997). Imaging and nano-manipulation of single biomolecules. Biophys Chem *68*, 63-72.

Funatsu, T., Harada, Y., Tokunada, M., Saito, K., and Yanagida, T. (1995). Imaging of single fluorescent molecules and individual ATP turnovers by single myosin molecules in aqueous solutions. Nature *374*, 555-559.

Geladopoulos, T. P., Sotiroudis, T. G., and Evangelopoulos, A. E. (1991). A Malachite Green Colorimetric Assay for Protein Phosphate Activity. Anal Biochem *192*, 112-116.

Gilbert, S. P., Moyer, M. L., and Johnson, K. A. (1998). Alternating site mechanism of the kinesin ATPase. Biochemistry *37*, 792-799.

Goldstein, L. S. (2001). Molecular motors: from one motor many tails to one motor many tales. Trends Cell Biol *11*, 477-482.

Grummt, M., Pistor, S., Lottspeich, F., and Schliwa, M. (1998a). Cloning and functional expression of a 'fast' fungal kinesin. FEBS Lett *427*, 79-84.

Grummt, M., Woehlke, G., Henningsen, U., Fuchs, S., Schleicher, M., and Schliwa, M. (1998b). Importance of a flexible hinge near the motor domain in kinesin-driven motility. Embo Journal *17*, 5536-5542.

Hackney, D. (1994a). Evidence for alternating head catalysis by kinesin during microtubule-stimulated ATP hydrolysis. PNAS *91*, 6865-6869.

Hackney, D., Levitt, J., and Suhan, J. (1992). Kinesin undergoes a 9 S to 6 S conformational transition. JBC *267*, 8696-8701.

Hackney, D. D. (1994b). The rate-limiting step in microtubule-stimulated ATP hydrolysis by dimeric kinesin head domains occurs while bound to the microtubule. J Biol Chem *269*, 16508-16511.

Hackney, D. D. (1995). Highly processive microtubule-stimulated ATP hydrolysis by dimeric kinesin head domains. Nature *377*, 448-450.

Hackney, D. D. (2002). Pathway of ADP-stimulated ADP release and dissociation of tethered kinesin from microtubules. Implications for the extent of processivity. Biochemistry *41*, 4437-4446.

Hackney, D. D., and F., S. M. (2000). Kinesin's IAK tail domain inhibits initial microtubule-stimulated ADP release. Nature Cell Biol *2*, 257-260.

Hackney, D. D., Stock, M. F., Moore, J., and Patterson, R. A. (2003). Modulation of kinesin half-site ADP release and kinetic processivity by a spacer between the head groups. Biochemistry *42*, 12011-12018.

Hahlen, K. (2004a) Crosslinking studies on the conventional kinesin of Neurospora crassa, Ludwig-Maxilians-Universität, München.

Hahlen, K., Mergler, J., Reinder, J., Sickmann, A. and Woehlke, G (2004b). The docked kinesin neck-linker induces a weak microtubule binding state. manuscript submitted.

Hancock, W. O., and Howard, J. (1998). Processivity of the motor protein kinesin requires two heads. J Cell Biol *140*, 1395-1405.

Hancock, W. O., and Howard, J. (1999). Kinesin's processivity results from mechanical and chemical coordination between the ATP hydrolysis cycles of the two motor domains. Proc Natl Acad Sci U S A *96*, 13147-13152.

Harrison, A., and King, S. M. (2000). The molecular anatomy of dynein. Essays Biochem *35*, 75-87.

Henningsen, U., and Schliwa, M. (1997). Reversal in the direction of movement of a molecular motor [see comments]. Nature *389*, 93-96 issn: 0028-0836.

Hirokawa, N. (1998). Kinesin and dynein superfamily proteins and the mechanism of organelle transport. Science *279*, 519-526.

Hirokawa, N., Pfister, K. K., Yorifuji, H., Wagner, M. C., Brady, S. T., and Bloom, G. S. (1989). Submolecular domains of bovine brain kinesin identified by electron microscopy and monoclonal antibody decoration. Cell *56*, 867-878.

Hirose, K., Lockhart, A., Cross, R. A., and Amos, L. A. (1996). Three-dimensional cryoelectron microscopy of dimeric kinesin and ncd motor domains on microtubules. Proc-Natl-Acad-Sci-U-S-A *93*, 9539-9544 issn: 0027-8424.

Hoenger, A., Sack, S., Thormahlen, M., Marx, A., Muller, J., Gross, H., and Mandelkow, E. (1998). Image reconstructions of microtubules decorated with monomeric and dimeric kinesins: comparison with x-ray structure and implications for motility. J Cell Biol *141*, 419-430.

Hoenger, A., Thormahlen, M., Diaz-Avalos, R., Doerhoefer, M., Goldie, K. N., Müller, J., and Mandelkow, E. (2000). A new look at the microtubule binding patterns of dimeric kinesins. J Mol Biol *297*, 1087-1103.

Hua, W., Chung, J., and Gelles, J. (2002). Distinguishing inchworm and hand-over-hand processive kinesin movement by neck rotation measurements. Science *295*, 844-848.

Huang, T., and Hackney, D. (1994a). Drosophila kinesin minimal motor domain expressed in Escherichia coli. Purification and kinetic characterization. JBC *269*, 16493-16501.

Huang, T., Suhan, J., and Hackney, D. (1994b). Drosophila kinesin motor domain extending to amino acid position 392 is dimeric when expressed in Escherichia coli. JBC *269*, 16502-16507.

Hunt, A. J., and Howard, J. (1993). Kinesin swivels to permit microtubule movement in any direction. Proc Natl Acad Sci U S A *90*, 11653-11657.

Hunter, A. W., Caplow, M., Coy, D. L., Hancock, W. O., Diez, S., Wordeman, L., and Howard, J. (2003). The kinesin-related protein MCAK is a microtubule depolymerase that forms an ATP-hydrolyzing complex at microtubule ends. Mol Cell *11*, 445-457.

Inoue, K., Akita, N., Yamashita, S., Shiba, T., and Fujita, T. (1990). Constitutive and inducible expression of a transgene directed by heterologous promoters in a trout liver cell line. Biochem Biophys Res Commun *173*, 1311-1316.

Inoue, Y., Toyoshima, Y. Y., Iwane, A. H., Morimoto, S., Higuchi, H., and Yanagida, T. (1997). Movements of truncated kinesin fragments with a short or an artificial flexible neck. Proc Natl Acad Sci U S A *94*, 7275-7280.

Jaud, J. (2003) Optische und mechanische Charakterisierung einzelner molekularer Motoren, Universität München, München.

Jiang, W., and Hackney, D. D. (1997a). Monomeric kinesin head domains hydrolyze multiple ATP molecules before release from a microtubule. J Biol Chem *272*, 5616-5621.

Jiang, W., Stock, M. F., Li, X., and Hackney, D. D. (1997b). Influence of the kinesin neck domain on dimerization and ATPase kinetics. J Biol Chem 272, 7626-7632.

Kallipolitou, A. (2002) Charakterisierung von Domänen des schnellen konventionellen Kinesins aus Neurospora crassa mit Hilfe C-terminal verkürzter und chimärer Mutanten, Doctoral thesis, Ludwig-Maximilians Universität München, München.

Kallipolitou, A., Deluca, D., Majdic, U., Lakamper, S., Cross, R., Meyhofer, E., Moroder, L., Schliwa, M., and Woehlke, G. (2001). Unusual properties of the fungal conventional kinesin neck domain from Neurospora crassa. Embo J 20, 6226-6235.

Kammerer, R. A., Schulthess, T., Landwehr, R., Lustig, A., Engel, J., Aebi, U., and Steinmetz, M. O. (1998). An autonomous folding unit mediates the assembly of two-stranded coiled coils. Proc Natl Acad Sci U S A 95, 13419-13424.

Kaseda, K., Higuchi, H., and Hirose, K. (2003). Alternate fast and slow stepping of a heterodimeric kinesin molecule. Nat Cell Biol 5, 1079-1082.

Kirchner, J., Seiler, S., Fuchs, S., and Schliwa, M. (1999a). Functional anatomy of the kinesin molecule in vivo. EMBO-J 18, 4404-4413 FTXT: Full Text ISSN: 0261-4189.

Kirchner, J., Woehlke, G., and Schliwa, M. (1999b). Universal and unique features of kinesin motors: insights from a comparison of fungal and animal conventional kinesins. Biol-Chem 380, 915-921 issn: 1431-6730.

Kozielski, F., Sack, S., Marx, A., Thormahlen, M., Schonbrunn, E., Biou, V., Thompson, A., Mandelkow, E. M., and Mandelkow, E. (1997). The crystal structure of dimeric kinesin and implications for microtubule-dependent motility. Cell 91 (7), 985-994.

Kull, F. J., and Endow, S. A. (2002). Kinesin: switch I & II and the motor mechanism. J Cell Sci 115, 15-23.

Kull, F. J., Sablin, E. P., Lau, R., Fletterick, R. J., and Vale, R. D. (1996). Crystal structure of the kinesin motor domain reveals a structural similarity to myosin. Nature 380, 550-555.

Kull, F. J., Vale, R. D., and Fletterick, R. J. (1998). The case for a common ancestor: kinesin and myosin motor proteins and G proteins. J Muscle Res Cell Motil 19, 877-886.

Laemmli, U. K. (1970). Cleavage of structural proteins during the assembly of the head of bacteriophage T4. Nature 227, 680-685.

Lakamper, S., Kallipolitou, A., Woehlke, G., Schliwa, M., and Meyhofer, E. (2003). Single fungal kinesin motor molecules move processively along microtubules. Biophys J 84, 1833-1843.

Lawrence, C. J., Dawe R.K., Christie, K. R. et al. (2004). A standardized kinesin nomenclature. accepted to Journal of Biological Chemistry.

Lehmler, C., Steinberg, G., Snetselaar, K. M., Schliwa, M., Kahmann, R., and Bolker, M. (1997). Identification of a motor protein required for filamentous growth in Ustilago maydis. Embo J 16, 3464-3473.

Ma, Y. Z., and Taylor, E. W. (1995a). Kinetic mechanism of kinesin motor domain. Biochemistry 34, 13233-13241.

Ma, Y. Z., and Taylor, E. W. (1995b). Mechanism of microtubule kinesin ATPase. Biochemistry 34, 13242-13251.

Ma, Y. Z., and Taylor, E. W. (1997a). Interacting head mechanism of microtubule-kinesin ATPase. J Biol Chem 272, 724-730.

Ma, Y. Z., and Taylor, E. W. (1997b). Kinetic mechanism of a monomeric kinesin construct. J Biol Chem 272, 717-723.

Mandelkow, E., and Johnson, K. A. (1998). The structural and mechanochemical cycle of kinesin. Trends-Biochem-Sci 23, 429-433 issn: 0167-7640.

Mandelkow, E.-M., Herrmann, M., and Rühl, U. (1985). Tubulin Domains Probed by Limited Proteolysis and Subunit-specific Antibodies. J Mol Biol 185, 311-327.

Mazumdar, M., and Cross, R. A. (1998). Engineering a lever into the kinesin neck. J Biol Chem 273, 29352-29359.

Meyhofer, E., and Howard, J. (1995). The force generated by a single kinesin molecule against an elastic load. Proc Natl Acad Sci U S A 92, 574-578.

Morii, H., Takenawa, T., Arisaka, F., and Shimizu, T. (1997). Identification of kinesin neck region as a stable alpha-helical coiled coil and its thermodynamic characterization. Biochemistry 36, 1933-1942.

Moyer, M. L., Gilbert, S. P., and Johnson, K. A. (1998). Pathway of ATP hydrolysis by monomeric and dimeric kinesin. Biochemistry 37, 800-813.

Nishiyama, M., Muto, E., Inoue, Y., Yanagida, T., and Higuchi, H. (2001). Substeps within the 8-nm step of the ATPase cycle of single kinesin molecules. Nat Cell Biol 3, 425-428.

Ovechkina, Y., and Wordeman, L. (2003). Unconventional motoring: an overview of the Kin C and Kin I kinesins. Traffic 4, 367-375.

Paschal, B. M., Holzbaur, E. L., Pfister, K. K., Clark, S., Meyer, D. I., and Vallee, R. B. (1993). Characterization of a 50-kDa polypeptide in cytoplasmic dynein preparations reveals a complex with p150GLUED and a novel actin. J Biol Chem 268, 15318-15323.

Requena, N., Alberti-Segui, C., Winzenburg, E., Horn, C., Schliwa, M., Philippsen, P., Liese, R., and Fischer, R. (2001). Genetic evidence for a microtubule-destabilizing effect of conventional kinesin and analysis of its consequences for the control of nuclear distribution in Aspergillus nidulans. Mol Microbiol 42, 121-132.

Rice, S., Cui, Y., Sindelar, C., Naber, N., Matuska, M., Vale, R., and Cooke, R. (2003). Thermodynamic properties of the kinesin neck-region docking to the catalytic core. Biophys J 84, 1844-1854.

Rice, S., Lin, A. W., Safer, D., Hart, C. L., Naber, N., Carragher, B. O., Cain, S. M., Pechatnikova, E., Wilson-Kubalek, E. M., Whittaker, M., *et al.* (1999). A structural change in the kinesin motor protein that drives motility. Nature *402*, 778-784.

Riddles, P. W., Blakeley, R. L., and Zerner, B. (1983). Reassessment of Ellman's reagent. Methods Enzymol *91*, 49-60.

Romberg, L., Pierce, D. W., and Vale, R. D. (1998). Role of the kinesin neck region in processive microtubule-based motility. J Cell Biol *140*, 1407-1416.

Rosenfeld, S. S., Fordyce, P. M., Jefferson, G. M., King, P. H., and Block, S. M. (2003). Stepping and stretching. How kinesin uses internal strain to walk processively. J Biol Chem *278*, 18550-18556.

Sablin, E. P., Case, R. B., Dai, S. C., Hart, C. L., Ruby, A., Vale, R. D., and Fletterick, R. J. (1998). Direction determination in the minus-end-directed kinesin motor ncd. Nature *395*, 813-816.

Sablin, E. P., and Fletterick, R. J. (2004). Coordination between motor domains in processive kinesins. J Biol Chem *279*, 15707-15710.

Sack, S., Muller, J., Marx, A., Thormahlen, M., Mandelkow, E. M., Brady, S. T., and Mandelkow, E. (1997). X-ray structure of motor and neck domains from rat brain kinesin. Biochemistry *36*, 16155-16165 issn: 10006-12960.

Schief, W. R., and Howard, J. (2001). Conformational changes during kinesin motility. Curr Opin Cell Biol *13*, 19-28.

Schliwa, M. (1989). Head and tail. Cell *56*, 719-720.

Schliwa, M. (2003). Molecular Motors (Weinheim, Wiley-VCH).

Schliwa, M., and Woehlke, G. (2003). Molecular motors. Nature *422*, 759-765.

Schnapp, B. J. (2003). Trafficking of signaling modules by kinesin motors. J Cell Sci *116*, 2125-2135.

Schnitzer, M. J., Visscher, K., and Block, S. M. (2000). Force production by single kinesin motors. Nat Cell Biol *2*, 718-723.

Schoch, C. L., Aist, J. R., Yoder, O. C., and Gillian Turgeon, B. (2003). A complete inventory of fungal kinesins in representative filamentous ascomycetes. Fungal Genet Biol *39*, 1-15.

Seeberger, C., Mandelkow, E., and Meyer, B. (2000). Conformational preferences of a synthetic 30mer peptide from the interface between the neck and stalk regions of kinesin. Biochemistry *39*, 12558-12567.

Seiler, S., Kirchner, J., Horn, C., Kallipolitou, A., Woehlke, G., and Schliwa, M. (2000). Cargo binding and regulatory sites in the tail of fungal conventional kinesin. Nature Cell Biol *2*, 333-338.

Seiler, S., Nargang, F. E., Steinberg, G., and Schliwa, M. (1997). Kinesin is essential for cell morphogenesis and polarized secretion in Neurospora crassa. Embo J *16*, 3025-3034.

Seiler, S., Plamann, M., and Schliwa, M. (1999). Kinesin and dynein mutants provide novel insights into the roles of vesicle traffic during cell morphogenesis in Neurospora. Curr Biol *9*, 779-785.

Sharp, D. J., Rogers, G. C., and Scholey, J. M. (2000). Microtubule motors in mitosis. Nature *407*, 41-47.

Sindelar, C. V., Budny, M. J., Rice, S., Naber, N., Fletterick, R., and Cooke, R. (2002). Two conformations in the human kinesin power stroke defined by X-ray crystallography and EPR spectroscopy. Nat Struct Biol *9*, 844-848.

Skiniotis, G., Surrey, T., Altmann, S., Gross, H., Song, Y. H., Mandelkow, E., and Hoenger, A. (2003). Nucleotide-induced conformations in the neck region of dimeric kinesin. Embo J *22*, 1518-1528.

Song, Y. H., Marx, A., Muller, J., Woehlke, G., Schliwa, M., Krebs, A., Hoenger, A., and Mandelkow, E. (2001). Structure of a fast kinesin: implications for ATPase mechanism and interactions with microtubules. Embo J *20*, 6213-6225.

Steinberg, G. (2000). The cellular roles of molecular motors in fungi. Trends Microbiol *8*, 162-168.

Steinberg, G., and Schliwa, M. (1995). The Neurospora organelle motor: a distant relative of conventional kinesin with unconventional properties. Mol Biol Cell *6*, 1605-1618.

Steinberg, G., and Schliwa, M. (1996). Characterization of the biophysical and motility properties of kinesin from the fungus Neurospora crassa. J Biol Chem *271*, 7516-7521.

Steinmetz, M. O., Stock, A., Schulthess, T., Landwehr, R., Lustig, A., Faix, J., Gerisch, G., Aebi, U., and Kammerer, R. A. (1998). A distinct 14 residue site triggers coiled-coil formation in cortexillin I. Embo J *17*, 1883-1891.

Stock, M. F., Guerrero, J., Cobb, B., Eggers, C. T., Huang, T. G., Li, X., and Hackney, D. D. (1999). Formation of the compact conformer of kinesin requires a COOH-terminal heavy chain domain and inhibits microtubule-stimulated ATPase activity. J-Biol-Chem *274*, 14617-14623 issn: 10021-19258.

Su, J. Y., Hodges, R. S., and Kay, C. M. (1994). Effect of chain length on the formation and stability of synthetic alpha-helical coiled coils. Biochemistry *33*, 15501-15510.

Svoboda, K., and Block, S. M. (1994). Force and velocity measured for single kinesin molecules. Cell *77*, 773-784.

Svoboda, K., Schmidt, C. F., Schnapp, B. J., and Block, S. M. (1993). Direct observation of kinesin stepping by optical trapping interferometry [see comments]. Nature *365*, 721-727.

Thormahlen, M., Marx, A., Sack, S., and Mandelkow, E. (1998). The coiled-coil helix in the neck of kinesin. Journal Of Structural Biology *122*, 30-41.

Thorn, K. S., Ubersax, J. A., and Vale, R. D. (2000). Engineering the processive run length of the kinesin motor. J Cell Biol *151*, 1093-1100.

Tomishige, M., and Vale, R. D. (2000). Controlling kinesin by reversible disulfide cross-linking. Identifying the motility-producing conformational change. J Cell Biol *151*, 1081-1092.

Tripet, B., and Hodges, R. S. (2002). Helix capping interactions stabilize the N-terminus of the kinesin neck coiled-coil. J Struct Biol *137*, 220-235.

Tripet, B., Vale, R. D., and Hodges, R. S. (1997). Demonstration of coiled-coil interactions within the kinesin neck region using synthetic peptides. Implications for motor activity. J Biol Chem *272*, 8946-8956.

Vale, R., Funatsu, T., Pierce, D., Romberg, L., Harada, Y., and Yanagida, T. (1996a). Direct Observation of Single Kinesin Molecules Moving Along Microtubules. Nature *380*, 451-453.

Vale, R. D., and Fletterick, R. J. (1997). The design plan of kinesin motors. Ann Rev Cell Dev Biol *13*, 745-777.

Vale, R. D., Funatsu, T., Pierce, D. W., Romberg, L., Harada, Y., and Yanagida, T. (1996b). Direct observation of single kinesin molecules moving along microtubules. Nature *380*, 451-453 issn: 0028-0836.

Vale, R. D., and Milligan, R. A. (2000). The way things move: looking under the hood of molecular motor proteins. Science *288*, 88-95.

Vale, R. D., Reese, T. S., and Sheetz, M. P. (1985). Identification of a novel force-generating protein, kinesin, involved in microtubule-based motility. Cell *42*, 39-50.

Verhey, K. J., Lizotte, D. L., Abramson, T., Barenboim, L., Schnapp, B. J., and Rapoport, T. A. (1998). Light chain-dependent regulation of Kinesin's interaction with microtubules. J-Cell-Biol *143*, 1053-1066 issn: 0021-9525.

Verhey, K. J., and Rapoport, T. A. (2001). Kinesin carries the signal. Trends Biochem Sci *26*, 545-550.

Vetter, I. R., and Wittinghofer, A. (2001). The guanine nucleotide-binding switch in three dimensions. Science *294*, 1299-1304.

Visscher, K., Schnitzer, M. J., and Block, S. M. (1999). Single kinesin molecules studied with a molecular force clamp. Nature *400*, 184-189.

Wang, Z., and Sheetz, M. P. (2000). The C-terminus of tubulin increases cytoplasmic dynein and kinesin processivity [In Process Citation]. Biophys J *78*, 1955-1964.

Woehlke, G. (2001). A look into kinesin's powerhouse. FEBS Lett *508*, 291-294.

Woehlke, G., and Schliwa, M. (2000). Wakling on Two Heads: the Many Talents of Kinesin. Nature Rev Mol Cell Biol *1*, 50-58.

Wu, Q., Sandrock, T. M., Turgeon, B. G., Yoder, O. C., Wirsel, S. G., and Aist, J. R. (1998). A fungal kinesin required for organelle motility, hyphal growth, and morphogenesis. Mol Biol Cell *9*, 89-101.

Xiang, X., and Fischer, R. (2004). Nuclear migration and positioning in filamentous fungi. Fungal Genet Biol *41*, 411-419.

Yildiz, A., Tomishige, M., Vale, R. D., and Selvin, P. R. (2004). Kinesin walks hand-over-hand. Science *303*, 676-678.

Young, E. C., Mahtani, H. K., and Gelles, J. (1998). One-headed kinesin derivatives move by a nonprocessive, low-duty ratio mechanism unlike that of two-headed kinesin. Biochemistry *37*, 3467-3479.

Zhou, N. E., Kay, C. M., and Hodges, R. S. (1993). Disulfide bond contribution to protein stability: positional effects of substitution in the hydrophobic core of the two-stranded alpha-helical coiled-coil. Biochemistry *32*, 3178-3187.

# Acknowledgements

I thank Prof. Dr. Manfred Schliwa for his guidance during my time in the institute, his interest in my work, the fruitful discussions and for sharing his ideas and experience.

I thank my supervisor Dr. Günther Woehlke for his support, for many helpful scientific suggestions and discussions and for sharing his technical expertise and experience.

I thank Prof. Marina Rodnina for advocating my thesis at the Faculty of Biosciences at the University of Witten/Herdecke and for her support during my studies and my PhD-work.

I thank Prof. M.A. Geeves for his kind willingness to review my thesis, for his support and for teaching me the secrets of enzyme kinetics during my time in his laboratory and the EMBO-course.

I thank Johann Jaud for sharing his great technical expertise, his interest and the patience that finally let the beads run. I also thank for litres of coffee, kilos of chocolate, excellent meals and the friendly atmosphere, that helped a lot to keep going after disappointments, and for having fun during days and nights with the laser trap.

I thank Prof. Dr. Matthias Rief for giving me the possibility to work with the optical laser trap in his institute and for helpful discussions and advices. I also thank Melanie Reisinger for her assistance with the single-molecule gliding assays.

I thank Dr. Katrin Hahlen for helpful assistance (especially with the microscope and the analysis of mysterious DNA-sequences), her company during hard times in the lab and her friendship.

I thank the former and current 3rd floor girls Athina Kallipolitou, Sarah Adio, Katrin Hahlen and Judith Mergler for the pleasant working atmosphere in the lab, the background music, many enjoyable tea and lunch times and for excellent diners.

I thank Sven Leier for great technical assistance, especially for his help to overcome the "Revco-GAU".

I thank Dr. Ralph Gräf, Dr. Günther Woehlke and Irene Schulz for helpful assistance with computer problems.

I thank all members of the ABiCB for the pleasant and stimulating atmosphere at work.

Ich danke meinen Eltern für ihre liebevolle Unterstützung und Förderung in allen Bereichen meines Lebens und ihre Anteilnahme an meinen Erfolgen und Enttäuschungen während meiner Doktorarbeit.

Ich danke meinem Mann Stephan für seinen Rat und sein Interesse an meiner Arbeit und für seine Geduld, sein liebevolles Verständnis, seine Ermutigungen und Begleitung.

Parts of this work that have been published elsewhere:

- Friederike Schaefer, D. Deluca, U. Majdic, J. Kirchner, M. Schliwa, L. Moroder and G. Woehlke (2003) EMBO J, Feb 3; 22 (3): 450-8
  <ins>A conserved tyrosine in the neck of a fungal kinesin regulates the catalytic motor core</ins>

- Friederike Bathe, K. Hahlen, R. Dombi, L. Driller and G. Woehlke (2005) Mol Biol Cell, Aug; 16 (8): 3529-37; Epub 2005 May 18
  <ins>The complex interplay between the neck and hinge domains in kinesin-1 dimerization and motor activity</ins>

- J. Jaud*, Friederike Bathe*, M. Rief, M. Schliwa and G. Woehlke (2006) Biophys J, Aug 15; 91(4):1407-12. Epub 2006 May 19
  <ins>Flexibility of the neck domain enhances Kinesin-1 motility under load</ins>
  *both authors contributed equally

The work presented here was carried out in the laboratory of Prof. Dr. Manfred Schliwa (Adolf-Butenandt-Institute of Cell Biology of the Ludwig-Maximilians-University, Munich) from October 2001 to October 2004. The work was supported by the Deutsche Forschungsgemeinschaft.

Die VDM Verlagsservicegesellschaft sucht für wissenschaftliche Verlage abgeschlossene und herausragende

## Dissertationen, Habilitationen, Diplomarbeiten, Master Theses, Magisterarbeiten usw.

für die kostenlose Publikation als Fachbuch.

Sie verfügen über eine Arbeit, die hohen inhaltlichen und formalen Ansprüchen genügt, und haben Interesse an einer honorarvergüteten Publikation?

Dann senden Sie bitte erste Informationen über sich und Ihre Arbeit per Email an *info@vdm-vsg.de*.

**Sie erhalten kurzfristig unser Feedback!**

VDM Verlagsservicegesellschaft mbH
Dudweiler Landstr. 99
D - 66123 Saarbrücken

Telefon  +49 681 3720 174
Fax       +49 681 3720 1749

**www.vdm-vsg.de**

Die VDM Verlagsservicegesellschaft mbH vertritt

Printed by Books on Demand GmbH, Norderstedt / Germany